MAR 0 9 1998   SCI

D1442966

7/02
4X   7/02

In Situ Treatment Technology

In Situ Treatment Technology

Evan K. Nyer   Sami Fam
Donald F. Kidd   Frank J. Johns II
Peter L. Palmer   Gary Boettcher
Tom L. Crossman   Suthan S. Suthersan

GERAGHTY & MILLER
ENVIRONMENTAL SCIENCE
AND ENGINEERING SERIES
a **heidemij** company

LEWIS PUBLISHERS

Boca Raton   New York   London   Tokyo

3 1336 04440 0878

Acquiring Editor:	Joel Stein
Cover Designer:	Denise Craig
Cover Concept:	Bill Cicio
Marketing Director:	Greg Daurelle
Prepress:	Kevin Luong
Senior Project Editor:	Andrea H. Demby
Manufacturing Assistant:	Sheri Schwartz

Library of Congress Cataloging-in-Publication Data

In situ treatment technology/Evan Nyer ... [et al.].
    p. cm.
   Includes bibliographical references and index.
   ISBN 0-87371-995-6 (alk. paper)
   1. In situ remediation.   I. Nyer, Evan K.
TD192.8.I5724   1996
  628.5'2--
dc20   95-49907
                                                      CIP

This book contains information obtained from authentic and highly regarded sources. Reprinted material is quoted with permission, and sources are indicated. A wide variety of references are listed. Reasonable efforts have been made to publish reliable data and information, but the author and the publisher cannot assume responsibility for the validity of all materials or for the consequences of their use.

Neither this book nor any part may be reproduced or transmitted in any form or by any means, electronic or mechanical, including photocopying, microfilming, and recording, or by any information storage and retrieval system, without prior permission in writing from the publisher.

CRC Press, Inc.'s consent does not extend to copying for general distribution, for promotion, for creating new works, or for resale. Specific permission must be obtained in writing from CRC Press for such copying.

Direct all inquiries to CRC Press, Inc., 2000 Corporate Blvd., N.W., Boca Raton, Florida 33431.

© 1996 by CRC Press, Inc.
Lewis Publishers is an imprint of CRC Press

No claim to original U.S. Government works
International Standard Book Number 0-87371-995-6
Library of Congress Card Number 95-49907
Printed in the United States of America 1 2 3 4 5 6 7 8 9 0
Printed on acid-free paper

Preface

The most exciting technical area in the remediation field today is "in place" or *in situ* technologies. The purpose of this book is to provide the reader with a single source that consolidates all of the information on the various *in situ* technologies. The main technology areas of bioremediation, vapor extraction, and sparging are discussed in individual chapters. This allows for an in-depth review of the state-of-the-art for each technology including laboratory and pilot plant studies, full-scale design, operation and maintenance, cost analysis, and case histories. New *in situ* technologies like fracturing, reactive walls, and vacuum enhanced recovery are also discussed in individual chapters. While the major subject areas are the same, pilot and full-scale designs, operations, costs, and case histories, the level of detail is lower due to less information available in the field. Emerging technologies like phytoremediation, temperature enhanced extraction, reactive zones, etc. are grouped together in a single chapter. Most of the work performed on these technologies have been at the pilot level with maybe one or two full-scale installations. Full-scale designs, operations, and cost data are very limited for these emerging technologies.

One chapter was required for non-*in situ* design considerations. Many of the *in situ* technologies use air movement as part of their applications. The air usually must be collected and brought above ground for treatment. Chapter 7 is devoted to discussing above-ground air treatment.

The book goes beyond discussing individual *in situ* technologies. The authors felt that it was very important for the reader to end up with an understanding of the geologic foundation and limitations of each of the technologies. The first chapter begins by explaining the limitations of pump-and-treat remediation. Designers have "progressed" to *in situ* technologies because the pump-and-treat remediations methods have failed to clean most sites. Chapter 1 provides the technical reasons that the pump-and-treat systems have had limited success, and how these same reasons may limit the success of *in situ* technologies. The information in Chapter 1 will also provide the reader with a basis to analyze and predict the possible success of any new *in situ* methods that are developed in the future.

I have tried to maintain the easy style of writing that my books normally enjoy. However, I felt that it was important to provide the reader the details necessary to be able to implement the *in situ* technologies. This dichotomy is one of the main reasons that I have ask the co-authors to participate in the book. Each of the co-authors works on a daily basis with the technology that they wrote about. I reviewed and rewrote each of the chapters, but they provided the meat. The result is, hopefully, a text that is still easy to read, but provides significant design and operational detail for each technology. The co-authors have their own bylines for the chapters that they wrote so that the reader will know the prime source of the information.

Many people have to be thanked beyond the co-authors. First, Geraghty & Miller has once again provided support and encouragement. There is no way that anyone could write a book today and put food on their table without the support of the firm that they work for. Geraghty & Miller has allowed me and the co-authors the time required for the book, and provided the staff support from drafting and secretaries. There are over 150 tables and figures in the book. In the beginning Jim Mallorey coordinated the production of the drawings. Brian Herman finished the coordination after I moved back to Florida. Both of these gentlemen put long hours into developing the drawings while trying to perform their normal workload. They were assisted by Bill Cicio, who provided the cover design concept, Eileen Shumacher, Deborah Quinlan, Christina Stewart, Steve Brown, and Dan Sohn. D'ann Dietrich provided secretarial support in Denver and Carla Gerstner finished the job once I moved back to Florida. In the technical area we have to thank David Schefer and Arul Ayyaswani. There is no way this book would have been finished without their support.

In situ technologies are an important part of being able to clean sites. I hope that the readers will find this book helpful in their applications of these new methodologies.

The Authors

 **Evan K. Nyer** is a Vice President with Geraghty & Miller, Inc., where he is responsible for maintaining and expanding the Company's technical expertise in geology/hydrogeology, engineering, modeling, risk assessment, and bioremediation. He has been active in the development of new treatment technologies for many years. His areas of research interest include biological treatment, suspended solids separation, chemical oxidation, *in situ* treatment, air treatment systems, and the application of new technologies to groundwater cleanup. He has been responsible for the strategies, technical design, and installation of more than 200 groundwater and soil remediation systems at contaminated sites throughout the United States.

Mr. Nyer received his graduate degree in environmental engineering from Purdue University and has authored two books, *Practical Techniques for Groundwater and Soil Remediation,* published by Lewis Publishers/CRC Press, Inc. and *Groundwater Treatment Technology,* now in its second edition, published by Van Nostrand Reinhold. Mr. Nyer is a regular contributor to *Groundwater Monitoring and Remediation,* having had his own column "Treatment Technology" in the periodical for the past seven years. He has presented numerous papers and conducted workshops for institutions including UCLA, the U.S. Environmental Protection Agency, the EPA Superfund Program, the University of Wisconsin, and the National Groundwater Association.

Peter L. Palmer, P.E., P.G., a registered professional engineer and professional geologist, has over 22 years of experience in conducting soil and groundwater contamination investigations, performing feasibility studies and alternatives assessments, and designing and implementing remediation programs. His projects have covered state as well as federal regulatory programs including RCRA and CERCLA. As Vice President of Engineering for Geraghty & Miller, Inc., he is responsible for fostering the quality and the growth of engineering services firmwide and administering the firm's innovative remedial technologies training and development program. He is involved in applications of these remedial technologies at sites located throughout the United States and lectures at seminars in both the United States and overseas in the integration of these technologies into remedial programs. Mr. Palmer is a member of the American Society of Civil Engineers, National Water Well Association, Water Environment Federation, and the National Society of Professional Engineers.

Suthan S. Suthersan, Ph.D., P.E., is a Vice President and Director of Remediation Technologies at Geraghty & Miller, Inc. His primary responsibilites include development and application of innovative *in situ* remediation technologies and provision of technical oversight on projects across the entire country. He has a Ph.D. in Environmental Engineering from the University of Toronto and is also a registered Professional Engineer in several states. His technology development efforts have been rewarded with many patents both awarded and pending. Dr. Suthersan has pioneered various conventional and modified applications of many *in situ* technologies such as *in situ* reactive walls, *in situ* air sparging, bioventing, *in situ* bioremediation, and pneumatic/hydraulic fracturing. His primary strength lies in developing the most cost-effective site-specific solutions utilizing cutting edge techniques. He has developed a national reputation for persuasion of the regulatory community to accept the most innovative remedial techniques.

Sami Fam, Ph.D., P.E., is President of Innovative Engineering Solutions, Inc. Dr. Fam has a Ph.D. and M.S. in Environmental Engineering from UCLA and a B.A. in Chemistry from Dartmouth College. He is a registered professional engineer in ten states. He is a published author in the areas of groundwater and soil remediation, chlorination by-products, oil and grease analysis, and urban runoff quality.

Frank J. Johns II, P.E., is a registered professional engineer and experienced designer of groundwater and soil remediation systems. Mr. Johns is a Technical Vice President at Geraghty & Miller, Inc. In this position, he is responsible for ensuring that projects not only meet the quality standards of the firm, but also that the technology selected is appropriate for the application. He also provides training for the staff to help keep them abreast of the advancing remedial technologies. Mr. Johns is a member of the Water Environment Federation, the American Water Works Association, and the American Society of Civil Engineers.

Donald F. Kidd, P.E., is a Senior Engineer at Geraghty & Miller's Phoenix, Arizona office. He has been with Geraghty & Miller for five-and-one-half years, with a total of eight years in environmental engineering consulting. Mr. Kidd specializes in the design and implementation of soil and groundwater remediation programs for orgnaic and inorganic contaminants. Mr. Kidd provides technical review for remediation engineering projects in many of Geraghty & Miller's offices throughout the western United States. Mr. Kidd obtained his B.S. in Geophysical Engineering from the Colarado School of Mines in 1985 and is a registered professional civil engineer in Arizona, California, Nevada, and Colorado.

Tom L. Crossman, Manager Bioremediation Services, has a B.S. in Dairy Technology/Microbiology. Mr. Crossman worked in the research and development of fermented food products, focused on accelerated aging and flavor-forming processes in cheeses, yogurts, etc., via enzymes and fermentation technologies.

He has applied biotechnology featuring immobilized enzyme and cell reactors resulting in two immobilized enzyme patents via controlled-porosity supports for bioreactor technology. At Geraghty & Miller, Inc., Mr. Crossman's primary focus is on *in situ* intrinsic bioattenuation and intrinsic reductive dechlorination, having evaluated and applied intrinsic remediation in over 60 remedial strategies. The recent focus in phytoremediation (use of vegetation) of groundwater and soils has led to current involvement in implementation of phytoremediation for remediation of TCE-contaminated groundwater and weathered hydrocarbons on vadose-zone soils. He is a member of Geraghty & Miller's National Technical Team and mentors intrinsic remediation and phytoremediation for the firm. He is the only consultant-member of the ACT 307 Subcommittee in the State of Michigan providing guidelines for bioremediation of chlorinateds and hydrocarbons.

Gary Boettcher has ten years of environmental experience obtained in the chemical industry, decontamination equipment manufacturing, hazardous waste treatment industry, and environmental consulting. He received a B.S. in microbiology in 1984 from the University of South Florida and completed graduate course work in public health. Mr. Boettcher specializes in investigation and remediation of impacted groundwater and soil. His technical expertise focuses on evaluation of physical, chemical, and biological remedial alternatives. Mr. Boettcher has been a project engineer, scientist, and manager on Federal and state Superfund, RCRA, and various industrial projects throughout the United States, including the Bahamas and Puerto Rico. Mr. Boettcher has developed added specialization in the area of bioremediation where he has designed, managed, and implemented *in situ* and *ex situ* bioremediation processes to treat hydrocarbons and industrial solvents. Mr. Boettcher has coauthored several papers on remediation and contributed to the development of U.S. Environmental Protection Agency (USEPA) guidance documents focusing on use of aerobic biological treatability studies at CERCLA sites.

Table of Contents

1   LIMITATIONS OF PUMP-AND-TREAT REMEDIATION METHODS

Evan K. Nyer

It has now become obvious that pump-and-treat methods of remediation will not be able to reach the required contaminant concentrations in most aquifers. Even with extended operating periods, we will not be able to call the contaminated aquifer "clean". Installing wells and forcing water to move past the contaminated portion of an aquifer will not remove the organic and heavy metal compounds at a high enough rate for the project to reach a conclusion in a reasonable amount of time.

Most people that have reached this conclusion have also decided that the next logical step is to treat the contaminants in place or *in situ*. The problem is that unless we have a complete understanding on why pump-and-treat systems cannot clean an aquifer, we will probably have the same problems when we go to design or evaluate an *in situ* method. Too often *in situ* methods are solely evaluated upon the rate at which they remove compounds from the aquifer. However, unless the new *in situ* method can overcome the limitations of the pump-and-treat, there may be limited reason for applying the technology. This means that we must not only analyze the rate at which the *in situ* method can remove contaminants from the aquifer, but also determine whether the new method can reach "clean" and, if so, at what cost.

Therefore, before we can review and discuss the various *in situ* methods, we must have a thorough understanding of the limitations of pump-and-treat. We have to go beyond the data analysis that shows that the process is limited. We have to understand the macro- and microprocesses that are occurring in the aquifer. Once the problem has been broken down into its components, then we can analyze an *in situ* method to see if it will overcome all of the individual problems that limit the effectiveness of pump-and-treat.

This chapter will break down the limitations of pump-and-treat into its basic components. We will first review water as the carrier of the contaminants and the creation of a contamination zone. Next, we will review how others have described

0-87371-995-6/96/$0.00+$.50
© 1996 by CRC Press, Inc.

the limitations of pump-and-treat. Finally, we will break down the limitations of pump-and-treat into component parts and analyze each one. These components will then become the basis for evaluating the various *in situ* methods.

WATER AS A CARRIER

One of the central points that must be made in this chapter is that water is the carrier when we are dealing with the aquifer. We drill a well into an aquifer, collect a water sample, and send the sample off to a laboratory for analysis to determine if the aquifer is contaminated. The contaminants found in this sample do not directly represent what is in the environment of the aquifer. The concentration of contaminants found in the water sample represents the relationship among the organic compounds located in the aquifer, their adsorption properties in relation to the aquifer soil, and their solubility in water. If we lose the same amount of acetone and decane to an aquifer, our groundwater sample will show more acetone then decane. The decane will adsorb to the soils in the aquifer and be less soluble in water than the acetone. The water will carry more acetone than decane to the sampling well. The concentration of acetone will be higher even though the same amount of acetone and decane are located in the aquifer.

Everything we do in the aquifer relates to water. It is an integral part of the aquifer environment. We have come to rely on the water in the aquifer (groundwater) as our measuring device and as our method of cleaning (pump-and-treat). Water is also the reason that the contaminants spread from their original release point. The best place to start our understanding of water as the carrier is to review how the contamination plume was originally formed.

THE CONTAMINATION PLUME

Plumes are created when a contaminant comes into contact with the aquifer. Contaminants can be released at the ground surface, into the unsaturated zone, or directly into the aquifer. Figure 1 shows a contaminant that was released at the surface. The contaminant travels down through the soil by the force of gravity. The organic compounds will adsorb to the soil as they move through the unsaturated zone. Assuming that all of the organics are not sorbed and that the contaminants do not encounter an impermeable layer, the contaminants will eventually reach the aquifer. We will discuss the unsaturated zone when we review air as a carrier.

Three things can happen when the contaminants reach the aquifer. Figure 1 shows the contaminant directly entering the aquifer. This will occur if the contaminant is very soluble in water. Table 1 provides the solubilities for 50 organic compounds. These are the organic compounds that are most likely to be found at a contaminated site. This book will provide the other properties of these compounds as that property is discussed in the text. As can be seen in Table 1, acetone is very soluble in water. It is listed as 1×10^6 mg/L. This means that the concentration can reach one million parts per million. In other words, acetone has infinite solubility in water. In general, the entire ketone and alcohol families of organic compounds are very soluble in water. If any of them are released to the ground environment, their movement will be represented by Figure 1.

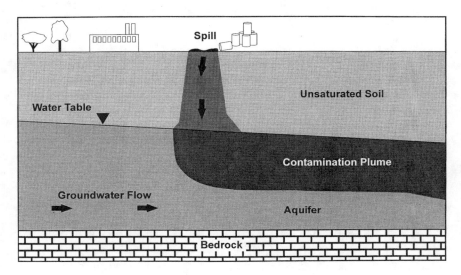

Figure 1 Contamination plume in an aquifer.

The rest of the organic compounds have a variety of solubilities. The values listed in Table 1 represent the maximum amount of organic that is soluble in water. Do not expect to find these concentrations in a groundwater sample, however. In fact, when the concentration in the groundwater sample reaches 5 to 10% of maximum solubility, then this is a strong indication that a source of pure compound is in close proximity to the sampling point (EPA report, 1992a, and Perry, 1984). (New research has shown that chlorinated hydrocarbon concentrations as low as 1% are a strong indication of pure compound.)

The solubility of the compound controls how much or at what rate the organic compound enters the groundwater of the aquifer. Therefore, the solubility also controls the mass rate at which the water can carry the organic away. From Table 1, 1 L of water can carry a maximum of 50 mg of hexachloroethane. However, the same liter of water can carry 14,000 mg of 2-hexanone. The actual movement of the organic compound in the groundwater is also affected by the affinity of the organic compound for sorption to the soil particles in the aquifer. This is quantified by the retardation factor of the compound and will be discussed later in this chapter.

Three things can happen when the organic compound has a relatively low solubility in water. If the organic liquid is lighter than water, then it will stop its downward movement when it reaches the aquifer. Figure 2 shows the movement of a lighter-than-water organic compound (gasoline, in the figure). If the organic liquid is heavier than water, it will continue its downward movement until it has been sorbed by the aquifer soil or it encounters an impermeable layer. In addition,

Table 1   Solubility for Specific Organic Compounds

	Compound	Solubility[a] (mg/L)	Ref.
1	Acenaphthene	3.42	2
2	Acetone	1×10^6 [a]	1
3	Aroclor 1254	1.2×10^{-2}	2
4	Benzene	1.75×10^3	1a
5	Benzo(a)pyrene	1.2×10^{-3}	2
6	Benzo(g,h,i)perylene	7×10^{-4}	2
7	Benzoic acid	2.7×10^3	2
8	Bromodichloromethane	4.4×10^3	2
9	Bromoform	3.01×10^3	1b
10	Carbon tetrachloride	7.57×10^2	1a
11	Chlorobenzene	4.66×10^2	1a
12	Chloroethane	5.74×10^3	2
13	Chloroform	8.2×10^3	1a
14	2-Chlorophenol	2.9×10^4	2
15	*p*-Dichlorobenzene (1,4)	7.9×10^1	2
16	1,1-Dichloroethane	5.5×10^3	1a
17	1,2-Dichloroethane	8.52×10^3	1a
18	1,1-Dichloroethylene	2.25×10^3	1a
19	*cis*-1,2-Dichloroethylene	3.5×10^3	1a
20	*trans*-1,2-Dichloroethylene	6.3×10^3	1a
21	2,4-Dichlorophenoxyacetic acid	6.2×10^2	2
22	Dimethyl phthalate	4.32×10^3	2
23	2,6-Dinitrotoluene	1.32×10^3	2
24	1,4-Dioxane	4.31×10^5	2
25	Ethylbenzene	1.52×10^2	1a
26	bis(2-Ethylhexy)phthalate	2.85×10^{-1}	2
27	Heptachlor	1.8×10^{-1}	2
28	Hexachlorobenzene	6×10^{-3}	1a
29	Hexachloroethane	5×10^1	2
30	2-Hexanone	1.4×10^4	2
31	Isophorone	1.2×10^4	2
32	Methylene chloride	2×10^4	1
33	Methyl ethyl ketone	2.68×10^5	1b
34	Methyl naphthalene	2.54×10^1	2a
35	Methyl *tert*-butyl ether	4.8	3
36	Naphthalene	3.2×10^1	2
37	Nitrobenzene	1.9×10^3	2
38	Pentachlorophenol	1.4×10^1	1
39	Phenol	9.3×10^4	1a,b
40	1,1,2,2-Tetrachloroethane	2.9×10^3	2
41	Tetrachloroethylene	1.5×10^2	1a
42	Tetrahydrofuran	3×10^{-1}	4
43	Toluene	5.35×10^2	1a
44	1,2,4-Trichlorobenzene	3×10^1	2
45	1,1,1-Trichloroethane	1.5×10^3	1a
46	1,1,2-Trichloroethane	4.5×10^3	1a
47	Trichloroethylene	1.1×10^3	1a
48	2,4,6-Trichlorophenol	8×10^2	2
49	Vinyl chloride	2.67×10^3	1a
50	*o*-Xylene	1.75×10^2	1c

[a] Solubility of 1,000,000 mg/L assigned because of reported "infinite solubility" in the literature.

Table 1   Solubility for Specific Organic Compounds

References:

1. *Superfund Public Health Evaluation Manual,* Office of Emergency and Remedial Response Office of Solid Waste and Emergency Response, U.S. Environmental Protection Agency, 1986.
    a. Environmental Criteria and Assessment Office (ECAO), U.S. Environmental Protection Agency, *Health Effects Assessments for Specific Chemicals,* 1985.
    b. Mabey, W.R., Smith, J.H., Rodoll, R.T., Johnson, H.L., Mill, T., Chou, T.W., Gates, J., Patridge, I.W., Jaber, H., and Vanderberg, D. Aquatic Fate Process Data for Organic Priority Pollutants, EPA Contract Nos. 68-01-3867 and 68-03-2981 by SRI International, for Monitoring and Data Support Division, Office of Water Regulations and Standards, Washington, D.C., 1982.
    c. Dawson et. al, *Physical/Chemical Properties of Hazardous Waste Constituents,* by Southeast Environmental Research Laboratory for U.S. Environmental Protection Agency, 1980.
2. EPA. Basics of Pump-and-Treat Groundwater Remediation Technology. Publication 600/8-901003, Robert S. Kerr Environmental Research Laboratory, U.S. Environmental Protection Agency, March 1990.
3. Texas Petrochemicals Corporation (manufacturer data), Gasoline Grade Methyl *tert*-Butyl Ether Shipping Specification and Technical Data, 1986.
4. Lide, D., Ed., *CRC Handbook of Chemistry and Physics,* 71st ed., Boca Raton, FL: CRC Press, 1990.

all compounds have some solubility in water, and a small amount of the organic compound will also be left in the aquifer as "residual saturation" as it moves down through the aquifer. The more soluble the organic compound, the more mass that will be lost to this process. Figure 3 shows the movement of a heavier-than-water organic compound (trichloroethylene, [TCE] in the figure). Basically, the organic compounds that are lighter than water will float on the water, and the organic compounds that are heavier than water will sink through the water.

Organic liquids found in the unsaturated zone or the aquifer are referred to as non-aqueous phase liquids, or NAPLs. When the NAPL is lighter than water, it is referred to as light NAPL, or LNAPL. When the NAPL is heavier than water it is referred to as dense NAPL, or DNAPL. These are all general terms and can be used for pure compounds or mixtures of organics. If the mixture of organic compounds has a resulting specific gravity greater than water, the entire mixture will sink through the water and can be referred to as a DNAPL. Even if a lighter-than-water organic is part of the mixture, it is the property of the entire mixture that will rule its vertical movement in the aquifer environment. For example, benzene is 5% of an organic mixture composed mainly of TCE. The specific gravity of the mixture is 1.25. Even though benzene is lighter than water, it will move with the entire mixture and sink through the water.

Table 2 summarizes the specific gravity for 50 organic compounds. In general, petroleum hydrocarbons are lighter than water, and chlorinated hydrocarbons are heavier than water. Table 2 does not include many petroleum hydrocarbons because

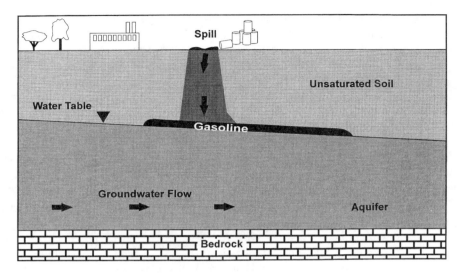

Figure 2   Gasoline spill encountering an aquifer.

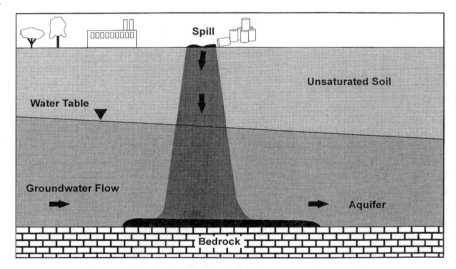

Figure 3   Trichloroethylene spill passing through an aquifer.

most of the organic compounds that make up the various petroleum products (gasoline, kerosene, fuel oil, etc.) are not hazardous. Table 3 provides the solubility and specific gravity for some of the organic compounds found in petroleum hydrocarbons.

As can be seen in Table 3, a compound such as decane is lighter than water with a specific gravity of 0.73. Decane is also not very soluble in water, with a solubility of 0.009 mg/L. If we had a release of pure decane to the ground, its travel pattern would look very similar to Figure 2; however, all petroleum products are a mixture of several organic compounds. Table 4 shows the concentration of the specific organic compounds, by volume, of three representative gasolines. The fate and transport of the mixture will be the result of the properties of the combined product and also the properties of the individual compounds. The specific gravity of the mixture will determine if the NAPL will float or sink as it encounters the aquifer. The solubility of the individual compounds will control how the compound dissolves into the groundwater of the aquifer and forms the plume. For example, gasoline is lighter than water and will float when it reaches the aquifer. The decane in the gasoline will mostly stay with the LNAPL. Benzene, however, is more soluble in water and will dissolve into the water and be part of the groundwater contamination plume. The same thing would happen in the case of our previous example, with benzene being a part of a DNAPL. The rate of release of the benzene to the water of the aquifer would relate to the individual compound and not be controlled by the mixture.

Figures 2 and 3 represent the movement of NAPLs that are not soluble in water. These drawings are a simplification of what happens in the real world. As can be seen in Table 1, all organic compounds have some solubility in water. Even hexachlorobenzene is slightly soluble in water. One liter of water will carry a maximum of 6 µg of hexachlorobenzene. Therefore, all of the NAPL that is in contact with the groundwater of the aquifer will release organic compounds to the water. Since groundwater is moving, the compounds will be carried away from the original point of contamination. Figures 4 and 5 show the plume of organics that are dissolved in the groundwater. These plumes are created from the interaction of NAPL with the moving groundwater.

There are several other ways in which a plume can form. For example, rain water can move through the vadose zone, dissolve contaminants that are located in that zone, and carry the contaminants to the aquifer. Volatile organic compounds in the vadose zone can volatilize and move in the gaseous phase (sometimes in a direction opposite to the direction of the groundwater flow). In this phase, the organics can come into contact with the aquifer and dissolve into the groundwater. There is no way to list all of the possible methods. The important point is that there is a source of organics, and when this source (all or part of it) comes into contact with the aquifer it will start to dissolve into the groundwater. The groundwater, which is moving, will carry the organics away from the original point of contact. The term "plume" refers to the dissolved phase of the organics in the aquifer.

PLUME MOVEMENT

While the objective of this book is to provide a detailed analysis of various *in situ* remedial technologies to clean the aquifer, to do so we must first understand how the plume moves in the aquifer. The main processes that affect plume movement are advection, dispersion, retardation, chemical precipitation, and biotransformation. Each compound will be affected differently by these processes.

Table 2   Specific Gravity for Specific Organic Compounds

	Compound	Specific[a] gravity	Ref.
1	Acenaphthene	1.069 (95°/95°)	1
2	Acetone	0.791	1
3	Aroclor 1254	1.5 (25°)	3
4	Benzene	0.87900	1
5	Benzo(a)pyrene	1.35 (25°)	4
6	Benzo(g,h,i)perylene	NA[b]	
7	Benzoic acid	1.316 (28°/4°)	1
8	Bromodichloromethane	2.006 (15°/4°)	1
9	Bromoform	2.903 (15°)	1
10	Carbon tetrachloride	1.594	1
11	Chlorobenzene	1.106	1
12	Chloroethane	.903 (10°)	1
13	Chloroform	1.49 (20°C liquid)	2
14	2-Chlorophenol	1.241 (18.2°/15°)	1
15	*p*-Dichlorobenzene (1,4)	1.458 (21°)	1
16	1,1-Dichloroethane	1.176	1
17	1,2-Dichloroethane	1.253	1
18	1,1-Dichloroethylene	1.250 (15°)	1
19	*cis*-1,2-Dichloroethylene	1.27 (25°C liquid)	2
20	*trans*-1,2-Dichloroethylene	1.27 (25°C liquid)	2
21	2,4-Dichlorophenoxyacetic acid	1.255	6
22	Dimethyl phthalate	1.189 (25°/25°)	1
23	2,6-Dinitrotoluene	1.283 (111°)	1
24	1,4-Dioxane	1.034	1
25	Ethylbenzene	.867	1
26	bis(2-Ethylhexy)phthalate	.984	1
27	Heptachlor	1.570	5
28	Hexachlorobenzene	2.044	1
29	Hexachloroethane	2.090	6
30	2-Hexanone	.815v (18°/4°)	1
31	Isophorone	.921 (25°)	2
32	Methylene chloride	1.366	1
33	Methyl ethyl ketone	.805	1
34	Methyl naphthalene	1.025 (14°/4°)	1
35	Methyl *tert*-butyl ether	.731	1
36	Naphthalene	1.145	1
37	Nitrobenzene	1.203	1
38	Pentachlorophenol	1.978 (22°)	1
39	Phenol	1.071 (25°/4°)	1
40	1,1,2,2-Tetrachloroethane	1.600	1
41	Tetrachloroethylene	1.631 (15°/4°)	1
42	Tetrahydrofuran	.8888 (21°/4°)	1
43	Toluene	.866	1
44	1,2,4-Trichlorobenzene	1.446 (26°)	1
45	1,1,1-Trichloroethane	1.346 (15°/4°)	
46	1,1,2-Trichloroethane	1.441 (25.5°/4°)	1
47	Trichloroethylene	1.466 (20°/20°)	1
48	2,4,6-Trichlorophenol	1.490 (75°/4°)	1
49	Vinyl Chloride	.908 (25°/25°)	1
50	*o*-Xylene	.880	1

[a] Specific gravity of compound at 20°C referred to water at 4°C (20°/4°) unless otherwise specified.
[b] NA = Not available.

References:

1. Dean, J.A. *Lange's Handbook of Chemistry,* 11th ed., New York: McGraw-Hill, 1973.

Table 2 Specific Gravity for Specific Organic Compounds

2. Weiss, G. *Hazardous Chemicals Data Book,* 2nd ed., New York: Noyes Data Corp., 1986.

3. "Draft Toxicological Profile for Selected PCBs," U.S. Public Health Service Agency for Toxic Substances and Disease Registry, November 1987.

4. "Draft Toxicological Profile for Benzo(a)pyrene," U.S. Public Health Service Agency for Toxic Substances and Disease Registry, October 1987.

5. Verschueren, K. *Handbook of Environmental Data on Organic Chemicals,* 2nd ed., New York: Van Nostrand Reinhold, 1983.

6. Merck Index, 9th ed., Rahway, NJ: Merck and Co., 1976.

Table 3 Physical/Chemical Properties of Selected Petroleum Hydrocarbons

Compound	Molecular weight	Specific gravity	Solubility mg/L (°C)	Boiling point (°C)	Vapor pressure at 1 atm and (°C)
Pentane	72.15	0.626	360 (16)	36	430 (20)
Hexane	86.17	0.66	13 (20)	68.7	120 (20)
Decane	142.28	0.73	0.009 (20)	173	2.7 (20)
Benzene	78.11	0.878	1780 (20)	80.1	76 (20)
Toluene	92.1	0.867	515 (20)	110.8	22 (20)
ortho-Xylene	106.17	0.88	175 (20)	144.4	5 (20)
meta-Xylene	106.17	0.86	175 (20)	139	6 (20)
para-Xylene	106.17	0.86	198 (25)	138.4	6.5 (20)

Note: Compiled from various sources.

Advection

Advection is the main process that moves the compounds in the aquifer. Advection is movement by bulk motion and is quantified by the value of the groundwater velocity. Under most conditions, ground water is constantly moving, although this movement is usually slow (typically 1 to 900 ft/y). To determine the flow and direction in an aquifer, basic information is needed. Once we collect or estimate that basic information, then the groundwater flow rate may be calculated. The relationship for flow is stated in Darcy's Law:

$$V = \frac{-K}{ne} \frac{\Delta h}{\Delta x}$$

where:

V = pore water velocity (L^3/T)

K = average hydraulic conductivity, a measure of the ability of the porous media to transmit water (L/T)

η = effective porosity for flow

$\dfrac{\Delta h}{\Delta x}$ = hydraulic gradient

Table 4   Some of the Major Constituents of the Gasoline Fraction (b.p. 36 to 117°C) in Selected Petroleums

Constituent	Volume (%)		
	Conroe, TX	Colinga, CA	Jennings, LA
Alkanes			
n-Pentane	0.33	0.44	1.12
n-Hexane	6.44	7.75	9.15
n-Heptane	6.90	5.94	8.42
2-Methylpentane	2.89	2.56	3.47
2,3-Dimethylhexane	0.22	1.30	2.39
Cycloalkanes			
Cyclopentane	0.96	1.76	0.67
Methylcyclopentane	6.51	10.29	5.01
Cyclohexane	10.40	7.63	7.13
Methylcyclohexane	22.00	14.55	18.07
Ethylcyclopentane	2.03	4.38	2.34
Trimethylcyclopentane	3.64	8.12	4.18
Aromatics			
Benzene	3.27	2.22	3.61
Toluene	16.19	7.94	12.02

Source: Adapted from Perry, J. J. *Petroleum Microbiology,* R. M. Atlas, Ed., New York: Macmillan, 1984.

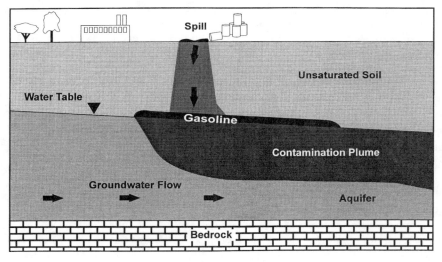

Figure 4 Contamination plume resulting from gasoline spill.

To determine the direction and velocity of flow, three or more wells may be drilled into the aquifer and the heads or water levels measured from a datum (typically mean sea level). Ground water will flow from high head to low head (the negative sign in Darcy's Law keeps the velocity positive as the gradient is always negative). The hydraulic conductivity (K) is a function of

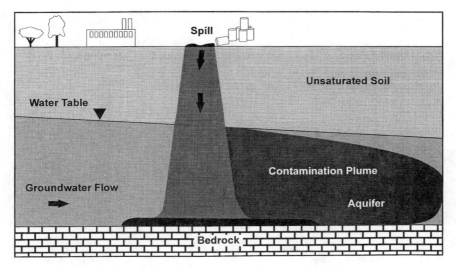

Figure 5 Contamination plume resulting from trichloroethylene spill.

the porous medium (aquifer) and the fluid (viscosity and specific weight); finer grained sediments such as silts and clays have relatively low values of K, whereas sand and gravel will have higher values. Other physical factors may affect the hydraulic conductivity including porosity, packing, sorting, solutioning, and fracturing.

This description refers to the perfect aquifer or a very large section of any aquifer. The problem is that when we look at water movement in small sections of the aquifer or the microenvironment, the water does not move with a uniform velocity. In the microenvironment, most aquifers have areas that have high water flow, areas that have low water flow, and areas that have no water flow. All aquifer soils are heterogeneous, some extremely so. The most obvious conditions that create extreme variable velocities are when the aquifer is constructed of material that forms large open areas. The best examples of this are fractured bedrock and karst geology. In both of these cases, the water flows mainly in the large, relatively open channels. Only the compounds that come into contact with one of these flow channels will combine with the water and form the plume. Since the flow channels cover only a small fraction of the cross-sectional area of the aquifer, the chances of the contaminant interacting with the water are greatly diminished. This limits the amount of organic that enters the water, but also limits the ability of the water to act as a carrier to clean the aquifer of the contaminant. In the field, the contaminant usually acts as a long-term source in these types of aquifers. The water movement in these aquifers can be very quick, and the plume length can be substantial.

Another problem that is created in these situations is in the investigation of the aquifer. We normally drill wells into the aquifer, take soil samples as we drill, and collect samples of the groundwater when the well is finished. In an aquifer in which the main flow is through open areas, if we do not encounter one of those areas, we will not be able to measure the contamination in the aquifer. When we consider the cross-sectional area represented by the well in relationship to the overall cross-sectional area of the aquifer, we realize the low probability of being able to take a representative sample of the contaminants in these types of aquifers.

The problem with using water as a carrier to clean these aquifers is very obvious, and very few systems have been designed that use pump-and-treat as the method to clean these aquifers. Pumping usually has been limited to controlling plume movement. It is beyond the scope of this book to discuss in detail these types of aquifers. Fractured bedrock and karst geology are unique situations. There are many fine books and articles written on the methods used to investigate contamination in these situations. The point that is being made here is that experts in the field have accepted that certain types of aquifers create flow patterns that obviously cause trouble when we try to measure contamination, predict plume movement, and clean the aquifer. The problem is that we do not recognize these same limitations when the same problems occur on a smaller scale in the aquifer. As stated before, most aquifers have areas in which the water is traveling at a high velocity, areas where water travels at a low velocity, and relatively stagnant areas. When we talk about advection, we must understand on a macro scale, and on a micro scale, how water travels through the aquifer. Let us review some of the other geology that creates flow patterns in the aquifers.

Many geologic situations can create zones of high and low velocity in an aquifer. The classic description of these two conditions are sand (or gravel) lenses and clay lenses. The soil particles in a sand lens are larger than the surrounding particles in the aquifer. While we are using the term "sand lens", we really are describing any aquifer that has an area in which the soil particles are larger than the surrounding particles, and the resultant permeability of the lens is higher than the permeability of the surrounding geology. In the real world, these areas are usually composed of coarse sands and gravel.

A typical sand lens is shown in Figure 6. As can be seen by the flow lines on Figure 6, the water flow in the aquifer will have a preference to move through the area of least resistance. The sand lens will allow most of the water flow in this area of the aquifer to travel through the sand. While the main path of the water in the aquifer may be the sand, the organic NAPLs may have a different path.

Figure 5 shows the contaminant traveling by gravity down through the aquifer. This flow path is perpendicular to the sand lens. The contaminant will form a plume as soon as it comes into contact with the water of the aquifer. However, while all of the water in the source area will be contaminated, the plume will move out from the source area because of the movement of the water. Under the microenvironment, this movement will be controlled by the sand lens. When the sand lens system is significant in the aquifer, then the flow pattern of the plume can be represented by Figure 7. The plume in Figure 7 has "fingers". These are areas of high flow that extend the plume faster along those paths.

Figure 7 is two dimensional. An aquifer is three dimensional. The same preferential flow paths can occur along the depth of the aquifer. Figure 8 shows the result of sand lenses along the depth of the aquifer. Once again, while all of the water is contaminated in the source area, the plume will mainly move by the

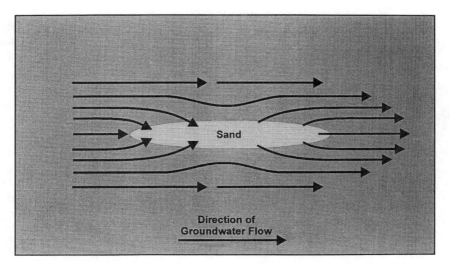

Figure 6 Groundwater flow pattern resulting from a sand lens.

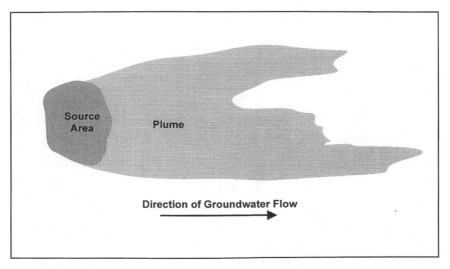

Figure 7 Sand lenses causing "fingers" in the plume, plan view.

movement of the water in the high flow areas of a sand lens. Figure 8 shows the importance of monitoring well screen placement in determining the contamination in the aquifer.

The opposite of the sand lens is the clay lens. When the soil material in the lens is smaller and the resultant permeability is less than that of the surrounding geology, the water will flow around that area. Figure 9 shows the water flow pattern when the aquifer includes a clay lens. While we refer to these areas as

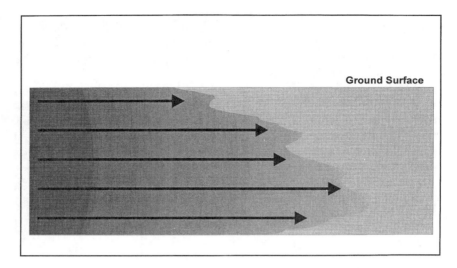

Figure 8 Sand lenses causing variations in flow patterns, section view.

clay lenses, they can be made up of any low-permeability soil material; however, they usually consist of clay or fine silt.

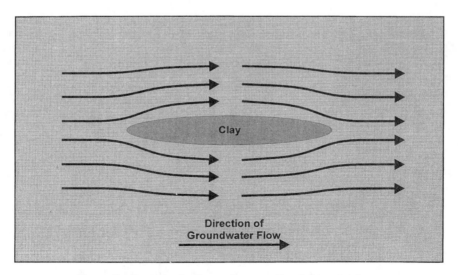

Figure 9 Groundwater flow pattern resulting from a clay lens.

The clay lens blocks water flow through an area of the aquifer. The clay lens may not be significant while the plume is spreading. Advection movement of the plume simply moves it around that area. However, the clay lens becomes very significant if it is contaminated and we are trying to remove contaminants from that area of the aquifer. Since water has a preference to move around the clay lens, then water will not act as a good carrier to remove the contaminants. The contaminants will remain in the clay lens and act as a source of contamination when we try to remediate the aquifer. We will discuss this further under diffusion in the next section of the chapter.

When we discuss plume movement, we usually only consider sand and clay lenses when they significantly affect the flow and pattern of the plume. When the sand lens system is large enough to create the fingers shown in Figures 7 and 8, then we include them in our description of the aquifer. However, if the sand lenses are relatively small, and other plume movement processes fill in the areas between the lenses, then, too often, we still refer to this type of aquifer as homogeneous.

This takes us back to the discussion on Darcy's Law. Darcy's Law is applicable to the perfect aquifer (homogeneous) or to a large enough section of any aquifer. The more heterogeneous factors in the aquifer, the larger the area that is needed to accurately describe the water movement by Darcy's Law and the average bulk hydraulic conductivity. When we have solution channels and/or bedrock fractures, a very large area is required. When we have sand and/or clay lenses, a relatively smaller area is required, depending upon the size of lenses.

The problem is that no aquifer is perfectly homogeneous. We can always select a small enough section of any aquifer so that the water movement in it cannot be described by Darcy's Law and the average hydraulic conductivity. Figure 10 shows a microenvironment of a homogeneous aquifer. The soil particles in this section are all of similar size. Even with the soil homogeneity, the small section of the aquifer has areas of high water velocity, areas that have low water velocity, and areas that are relatively stagnant. Szecsody described this situation by Figure 11 in 1989. (Szecsody and Bales, 1989).

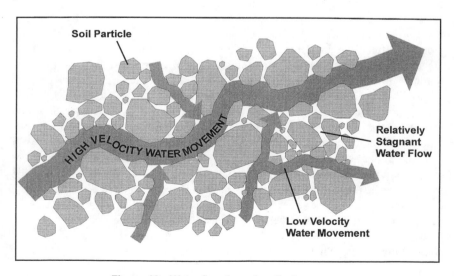

Figure 10 Water flow through soil microcosm.

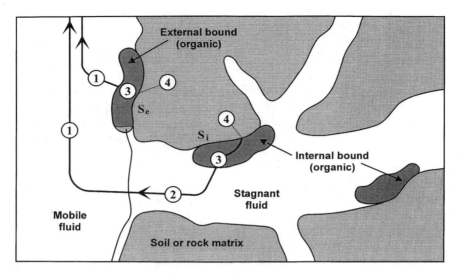

Figure 11 Conceptual model of sorption in porous aggregate.

This discussion is moot when we are describing plume movement. When we are describing how a plume moves in an aquifer, the micro-movement of the water does not matter. The plume moves mainly by advection, and Darcy's Law describes that advection. The microenvironment "fills in" by another method of organic movement, dispersion, which will be discussed in the next section of this chapter. Darcy's Law is dependent only upon selecting a large enough section of the aquifer. The small section that is described in Figure 10 is not significant in that analysis or the advection movement of the plume.

However, when we are trying to analyze water as a carrier to move contaminants out of an aquifer, then the micro analysis becomes very significant. During pump-and-treat, we are dependent upon the water coming into contact with the contaminants, picking up the contaminant (solubilizing), and carrying the contaminant to our recovery well. To completely clean an aquifer, the water must get to all of the soil particles. Figure 10 shows that the water will not flow past every particle in the aquifer. The contaminants located in the minor flow and relatively stagnant flow areas will have to be moved out of those areas by means other than water movement or advection.

The final questions become: How large is the area in the microenvironment that does not have water moving by each soil particle? While advection is the main force in plume movement, is it the main force when we try to remove contaminants from an aquifer? We need to discuss the other types of movement before we can fully answer these questions. The point to be made is that even though advection is the main process in plume movement, it may not be the controlling process when we go to remediate the plume.

Dispersion

Dispersion creates two main results in the movement of the plume. Both relate to a spreading of the contaminants. The first result can be shown in Figure 12. Additional spreading beyond the plume movement caused by advection can be the result of dispersion. Keely cautions that permeability differences between strata can be a main cause of plume spreading by dispersion (Keely, 1989). Figure 13 shows the difference between spreading by advection and spreading by dispersion.

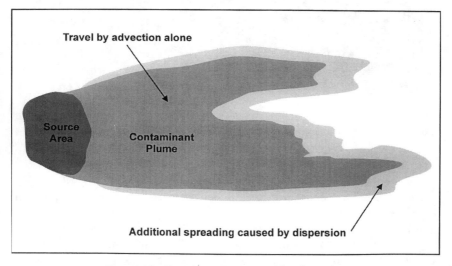

Figure 12 Containment movement by dispersion.

The second movement of the contaminants that can relate to dispersion is filling in the areas between main advection movement. As discussed in the previous section, the main movement of the plume is by advection, water movement. The trouble is that there are main flow paths and minor flow paths. The plume would mainly spread along these major flow paths and leave the minor flow paths clean. However, the real rate of flow of the water in an aquifer is very slow. Dispersion helps move the contaminants into the areas of relatively slow flow as the plume moves along by advection. Dispersion fills in the areas that do not have direct contact with the main water flow. Even when sand or clay lenses are present, dispersion fills in the area in between the areas of relatively high flow.

Dispersion can be divided into two components: mechanical movement and chemical movement. Mechanical dispersion comes from the difference in velocity from water molecules in contact with soil particles and water molecules moving in between the soil particles. When the water molecule is in contact with the soil particle or when it 'wets' the grain of sand, silt, or clay, its movement is restricted.

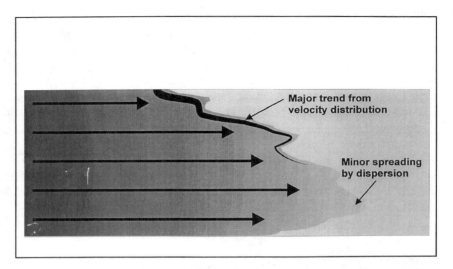

Figure 13   Cross-sectional view of contaminant spreading. Permeability differences between strata cause comparable differences in advection and, hence, plume spreading.

The molecule in contact will slow its movement relative to the rest of the water molecules. The molecules that move in between the soil particles and have no direct contact with the soil will move at a relatively higher rate. The difference in these velocities creates the mechanical component of dispersion.

The chemical component of dispersion is the result of molecular diffusion. All molecules have independent motion. While the main movement of a contaminant in water will be by the movement of the water carrying the compound, the compound will also have random movement within the water. Diffusion is a minor part of plume spreading. When the water is moving quickly, it has a negligible impact. However, when the plume movement is very slow, diffusion can cause significant spreading. What is more important from a remediation point of view is that diffusion can spread the contaminants into areas that have relatively slow flow or are completely stagnant even when the water is moving fast. This process will place contaminants in areas of the aquifer that have no direct water flow contact. This is not significant when we are watching or trying to predict the plume movement. However, when we reverse the process and use the water to bring back the contaminants, these compounds are out of contact with the flow of water. As the water in the main flow paths gets cleaner, the contaminants that have diffused into the low flow areas reverse course. The compounds naturally diffuse from an area of high concentration (the micro clay lens or other stagnant area) to an area of low concentration (the main water flow path). Thus, these micro areas become a source of low concentration contaminants as we use pump-and-treat to remediate a site.

Retardation

The next major influence on the movement of contaminants in an aquifer is retardation. Retardation is the result of the contaminants being attracted by and held to the surface of the aquifer solids. Sorption and ion exchange are the physical and chemical processes that work between the contaminants and the solids. The result is that as the water carries the contaminants through the aquifer, the contaminant comes into contact with the aquifer solids. The solids temporarily hold or retard the contaminant as the water continues to move. Thus the water moves faster than the contaminant. The difference in the rates relates to how "temporarily" the aquifer solids hold the contaminant. There are several factors that influence retardation. The properties of the contaminant that affect retardation can be approximated by measurements such as bulk density and partition coefficients. Table 5 provides the octanol-water partition coefficients (K_{ow}) for 50 organic compounds.

Generally, the less soluble and/or more complex the compound, the higher the attraction to the aquifer solids. Carbon tetrachloride has a higher octanol-water coefficient than trichlorethlene. It will have a higher attraction to the aquifer solids and, thus, move slower than TCE in the aquifer.

The properties of the aquifer solids also have a great effect on the retardation of the contaminants. In general, the higher the organic content that can be measured by total organic carbon (TOC) of the soil, the more it retards the movement of the contaminants. The ion exchange capacity of the aquifer solids can also affect contaminants that exist as ionic species, mainly heavy metals.

The only real way to account for all of these factors in an actual aquifer is by direct measurement. The laboratory can be used by taking soil samples from the site and passing the compounds through a test cell. While these data will be more accurate than the results predicted from partition coefficient and TOC, the best measurement of the effective mobility of each contaminant is made from observing the actual plume composition and its spreading over time.

Retardation is reported as a ratio of groundwater movement to the movement of the organic compounds, or the velocity of uncontaminated ground water divided by the velocity of the contaminant. The ground water is considered the base. If the contaminant travels at one half the speed of the ground water, then the retardation factor is 2. If the contaminant travels at one fourth the speed of the ground water, then the retardation factor is 4. Since both the contaminant properties and the aquifer solid properties all contribute to the retardation of a specific compound, we cannot provide a table of the retardation factors for each compound. Ranges can be provided for compounds and types of soil, but more precise numbers will have to be generated at each site.

One example of the combined effect of diffusion and retardation is the work by Roberts et al. (1986). Working on the aquifer at the Canadian Forces Base (Borden, Ontario, Canada), Roberts studied the movement of organic compounds over a 2-year period. They created a pulse of organics in the aquifer and then monitored the concentrations at three locations in the downstream aquifer. The

Table 5   Octanol Water Coefficients (K_{ow}) for Specific Organic Compounds

	Compound	K_{ow}	Ref.
1	Acenaphthene	1.0×10^4	2
2	Acetone	6×10^1	1d
3	Aroclor 1254	1.07×10^6	2
4	Benzene	1.3×10^2	1a
5	Benzo(a)pyrene	1.15×10^6	2
6	Benzo(g,h,i)perylene	3.24×10^6	2
7	Benzoic acid	7.4×10^1	2
8	Bromodichloromethane	7.6×10^1	2
9	Bromoform	2.5×10^2	1b
10	Carbon tetrachloride	4.4×10^2	1a
11	Chlorobenzene	6.9×10^2	1a
12	Chloroethane	3.5×10^1	2
13	Chloroform	9.3×10^1	1a
14	2-Chlorophenol	1.5×10^1	2
15	*p*-Dichlorobenzene (1,4)	3.9×10^3	2
16	1,1-Dichloroethane	6.2×10^1	1a
17	1,2-Dichloroethane	3.0×10^1	1a
18	1,1-Dichloroethylene	6.9×10^1	1a
19	*cis*-1,2-Dichloroethylene	5	1a
20	*trans*-1,2-Dichloroethylene	3	1a
21	2,4-Dichlorophenoxyacetic acid	6.5×10^2	2
22	Dimethyl phthalate	1.3×10^2	2
23	2,6-Dinitrotoluene	1.0×10^2	2
24	1,4-Dioxane	1.02	2
25	Ethylbenzene	1.4×10^3	1a
26	bis(2-Ethylhexy)phthalate	9.5×10^3	2
27	Heptachlor	2.51×10^4	2
28	Hexachlorobenzene	1.7×10^5	1a
29	Hexachloroethane	3.98×10^4	2
30	2-Hexanone	2.5×10^1	3
31	Isophorone	5.0×10^1	2
32	Methylene chloride	1.9×10^1	1b
33	Methyl ethyl ketone	1.8	1a
34	Methyl naphthalene	1.3×10^4	2
35	Methyl *tert*-butyl ether	NA[a]	
36	Naphthalene	2.8×10^3	2
37	Nitrobenzene	7.1×10^1	2
38	Pentachlorophenol	1.0×10^5	1b
39	Phenol	2.9×10^1	1a
40	1,1,2,2-Tetrachloroethane	2.5×10^2	2
41	Tetrachloroethylene	3.9×10^2	1a
42	Tetrahydrofuran	6.6	4
43	Toluene	1.3×10^2	1a
44	1,2,4-Trichlorobenzene	2.0×10^4	2
45	1,1,1-Trichloroethane	3.2×10^2	1b
46	1,1,2-Trichloroethane	2.9×10^2	1a
47	Trichloroethylene	2.4×10^2	1a
48	2,4,6-Trichlorophenol	7.4×10^1	2
49	Vinyl chloride	2.4×10^1	1a
50	*o*-Xylene	8.9×10^2	1c

[a] NA = not available.

Table 5 Octanol Water Coefficients (K_{ow}) for Specific Organic Compounds

References:

1. *Superfund Public Health Evaluation Manual,* Office of Emergency and Remedial Response Office of Solid Waste and Emergency Response, U.S. Environmental Protection Agency, 1986.
   a. Environmental Criteria and Assessment Office (ECAO), U.S. Environmental Protection Agency, *Health Effects Assessments for Specific Chemicals,* 1985.
   b. Mabey, W.R., Smith, J.H., Rodoll, R.T., Johnson, H.L., Mill, T., Chou, T.W., Gates, J., Patridge, I.W., Jaber, H., and Vanderberg, D., Aquatic Fate Process Data for Organic Priority Pollutants, EPA Contract Nos. 68-01-3867 and 68-03-2981 by SRI International, for Monitoring and Data Support Division, Office of Water Regulations and Standards, Washington, D.C., 1982.
   c. Dawson et. al, *Physical/Chemical Properties of Hazardous Waste Constituents,* by Southeast Environmental Research Laboratory for U.S. Environmental Protection Agency, 1980.
   d. Handbook of Environmental Data for Organic Chemicals, 2nd ed., New York: Van Nostrand Reinhold Co., 1983.
2. EPA. Basics of Pump-and-Treat Groundwater Remediation Technology, Publication 600/8-901003, Robert S. Kerr Environmental Research Laboratory, U.S. Environmental Protection Agency, March 1990.
3. Lyman, W.J. et al. "Research and Development of Methods for Estimating Physicochemical Properties of Organic Compounds of Environmental Concern, DAMD 17-78-C-8073," June 1981
4. EPA. Hazardous Waste Treatment, Storage and Disposal Facilities (TSDF) Air Emissions Model (draft document). Washington, D.C.: U.S. Environmental Protection Agency, April 1989.

results are shown in Figure 14. As can be seen, the tracer and organic compounds all have a tendency to tail off with time. The higher the retardation factor for the compound, the longer the tail.

The tails in this study were created in a homogeneous, sand aquifer over a 5-m sampling area. Due to the nature of the aquifer and the short distances, the effect from diffusion will be minor compared to retardation. As the aquifer becomes less homogenous, the tails of concentration will lengthen.

One final note before we leave retardation. The same properties that can slow the movement of contaminants in an aquifer can also combine to speed the movement of the contaminants. When we discuss aquifer solids, we usually mean relatively large soil particles. However, very small soil particles also exist in the aquifer, colloids. These small particles are not stationary but can move with the water. The water once again acts as a carrier for these solids. When a contaminant is attached to one of these small particles it travels with the particle. Instead of being retarded, it moves at the rate that the ground water is moving. Sampling methods become very important to understanding more than simply "how much"

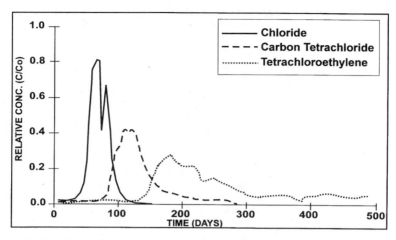

Figure 14   Breakthrough responses for chloride, carbon tetrachloride, and ethylene at well 4 (x = 5.0 m, y = 0.0 m, z = −3.26 m). (From Roberts, P.V., Goltz, M.N., and MacKay, D.M., *Water Resour. Res.,* vol. 22, no. 13, pp. 2047-2058, December 1986. With permission.)

compound is at a point in the aquifer. How the compound got to that point is also important in understanding what the data really mean.

Chemical Precipitation and Biotransformation

The last thing that can happen to the contaminants as they move through the aquifer is that they can disappear. The chemicals can change their form, or they can be completely destroyed. These actions, or reactions, can take place chemically or biochemically.

The best known chemical reaction in the aquifer is *precipitation.* Heavy metals in an ionic or dissolved phase can react with the aquifer soils or the ground water and precipitate. Once the metal is in a solid phase form, it no longer travels with the ground water (except if the precipitate is a colloid that can continue to be carried by the water). If the ground water does not carry the metal, then it is no longer a part of the plume, and it will not show up at any of the monitoring points.

The solubility of most heavy metals is related to the pH of the water. Figure 15 shows the solubility for several metals in relationship to the pH of the water. Two points must be made about these curves. First, they are only relative and do not represent exact values. Second, every water will be slightly different as far as specific solubility and the pH of minimum solubility.

The pH surrounding the immediate area where the dissolved metal comes into initial contact with the aquifer may be low. This could be the interaction of the rest of the chemicals associated with the heavy metal. When industry uses high concentrations of heavy metals, the solutions are usually very acidic. At the low pH, the metal would be relatively soluble and move with the groundwater movements. As the groundwater moved from the immediate area, the aquifer

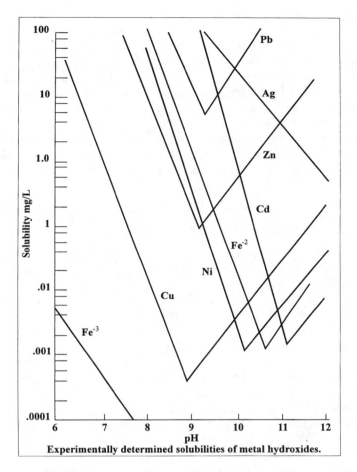

Figure 15 Solubilities of metal hydroxides at various pHs. (Courtesy of Graver Water, Union, NJ.)

solids would interact with the components of the groundwater and tend to neutralize the low pH. As the pH increased, the metal would precipitate as a result of natural water chemistry. The dissolved metal could also directly interact with the aquifer solids and precipitate.

The other major reaction in aquifers is *biochemical*. Bacteria are ubiquitous. They exist everywhere, including aquifers. The bacteria can also interact with the contaminants and change their form or completely destroy them. Chapter 3 will cover *in situ* biogeochemical reactions in detail. Bacteria can interact with metals, changing their valence state and making them more or less soluble. Iron fouling in wells is an example of bacterial interaction with metals.

A more universal interaction is between bacteria and organic contaminants. The bacteria can use many contaminants as food sources. The bacteria derive energy and building blocks for new bacteria from the carbon molecules in the contaminant. In addition, several of the enzymes that bacteria produce can interact with contaminants that are not considered food. The contaminants can be transformed into new chemicals or can be completely mineralized to CO_2. An example of transformation is that trichlorethane can be transformed into vinyl chloride in an aquifer. Petroleum hydrocarbons have a tendency for complete mineralization. In both cases, the original compound disappears. In the first case, a new compound is created. The new compound is also a contaminant, and its movement in the aquifer will be governed by its chemical properties.

Both chemical and biochemical reactions can change the contaminant. This has to be understood when interpreting the data and plume movement and is very important to understand when trying to use water as the carrier to remove the contaminant.

NON-AQUEOUS PHASE LIQUIDS — NAPL

The final area that we will discuss in this section of water as a carrier is the effect of NAPLs on water movement in the aquifer. Many papers and textbooks discuss NAPLs as a source of contamination; we have already discussed this aspect of NAPLs in a previous section. It is unusual to consider these pure compounds in a context of water movement. However, as you will see in the following discussion, the NAPLs also can have a significant effect on water as the carrier for our pump-and-treat remediations.

Most organic contaminants that we encounter in groundwater exist as liquids when they are in their pure form. Pure trichloroethylene, for example, is a free-flowing liquid. If it is lost to the ground, it will flow as a liquid through the ground and aquifer solids. As stated earlier, the liquid will sorb to the soil particles in both the unsaturated zone and the aquifer as it travels down through the ground.

Once the NAPL is in the unsaturated zone or the aquifer it can exist in several physical forms. First, the NAPL can coat the soil particles (Figure 16) or sorb to the soil particles. Second, the NAPL can actually fill all of the pore space between soil particles (Figure 17). Finally, the NAPL can fill fractures or voids in the subsurface materials.

It is not always easy to find the presence of NAPLs. The easiest method is direct evidence; LNAPLs are normally found by this method. When the well is drilled, we find the pure compound floating on top of the water either in measurable amounts or simply as an oil sheen on top of the groundwater. Direct evidence of DNAPLs is much harder to come by. The material sorbed to the soil does not always enter the water and show up in the monitoring well as a separate phase. Also, encountering where the actual DNAPL traveled down through the ground and trying to find it in core samples is not always reliable. Other methods have been developed to help determine if NAPLs are present in an aquifer. These methods are indirect measurements of the presence of NAPLs. The EPA recommends four indirect methods for determining NAPL presence: (1) high concentration of contaminants in the ground water, (2) depth of contamination in the

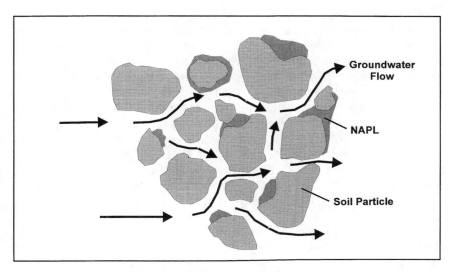

Figure 16 NAPLs coating soil particles.

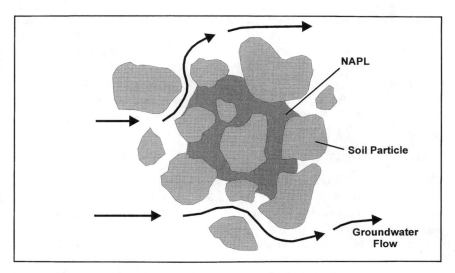

Figure 17 NAPLs completely filling in pores.

aquifer, (3) persistence of contaminants in a pump-and-treat system, and (4) contamination source characteristics. (EPA report, 1992a).

As can be seen by the above discussion, NAPLs can be a significant source of mass of contaminants in an aquifer. Since they are pure compounds, they have a high mass in a relatively small space. What is more significant when we try to remove these compounds is the amount of space that they occupy in the aquifer or vadose zone. When the NAPL coats the soil particle or completely fills the pore space between particles, the movement of other liquids in that same area is severely restricted. Water will not be able to move through an area where the pore spaces are completely filled with NAPL. Water will have severely restricted movement in areas where the NAPL coats the soil particle and reduces the space between soil particles. All of these restrictions force the water to move around the area where the NAPL is present.

As we discussed before, water must have intimate contact with the contaminants if it is to be used as a carrier to remove the contaminants from underground. The presence of NAPLs in the aquifer causes two problems. First, the water cannot come into direct contact with the NAPL because of movement restrictions. Second, the NAPL represents a huge source of contaminant. Once again, the only method for this contaminant to be removed from the aquifer with water as a carrier is for the contaminant to defuse over to an area where the water is able to move through the aquifer.

One final property should be discussed before we leave the NAPL discussion. NAPL and water movement through the ground are significantly different. We cannot assume that the main channels that allow water flow will be the same main channels that will help the NAPL flow through the ground. First, the NAPL will have significantly different chemical and physical properties from water. The interaction with the soil particles also will be very different. In addition, the NAPLs have a tendency to flow vertically in the soil with gravity being the main driving force in the downward movement. Water in the saturated zones will mainly flow horizontal. The main flow channels that exist horizontally will not be the same flow channels that exist vertically, although they will intersect. This means that significant sources of contamination can exist in the aquifer in areas that do not receive high flows of water. These NAPLs can then remain stagnant and serve as a source of contamination for years.

PUMP AND TREAT

Now that we have contaminated the aquifer, the objective is to put it back in its original condition or to clean the aquifer. The problem is that our main tool has severe limitations. On a purely logical basis, one would think that in a saturated zone, water would be everywhere and therefore would be in contact with all possible points of contamination in the aquifer. This is true; however, the problem occurs when we try to use the water to carry the contaminants out of the aquifer. All of the restrictions discussed above prevent the total removal of the contaminant. Let us review some real-world examples of this problem.

The best known and most complete study of the effectiveness of a pump-and-treat system was performed by the Environmental Protection Agency (EPA). Phase One of the study was completed in 1989 and covered 19 Superfund sites. In 1991, during Phase Two of the study, 5 more sites were included in addition to more data from the original 19 sites. The study is summarized in an EPA report titled, "Evaluation of Ground-Water Extraction Remedies: Phase II," February

1992, Publication 9355.4-05. The conclusions made from that study are as follows:

1. Data collected, both site characterization data prior to system design and subsequent operational data, were not sufficient to fully assess contaminant movement or groundwater system response to extraction.
2. In the majority of cases studies (15 of the 24 sites), the groundwater extraction systems were able to achieve hydraulic containment of the dissolved-phase contaminant plume.
3. Extraction systems were often able to remove a substantial mass of contamination from the aquifer.
4. When extraction systems were started up, contaminant concentrations usually showed a rapid initial decrease, but then tended to level off or decrease at a greatly reduced rate. This may be a result of the type of monitoring data collected as much as a reflection of an actual phenomenon of groundwater extraction systems. For example, it can reflect either successful remediation as the contaminated zone shrinks and less-contaminated groundwater is pulled into the extraction system, or poor placement of groundwater monitoring wells.
5. Based on the available information, potential NAPL presence was not addressed during site investigations at 14 of the 24 sites. At five sites they were "addressed" because they were encountered unexpectedly during the investigation. As a result, it is difficult to determine NAPL presence conclusively from available site data. Because NAPLs were not addressed in the site investigation, they also were not addressed in the remedial design. Consequently, a groundwater extraction system may be performing as designed (removing dissolved phase contaminants) even though it will not achieve the cleanup goals within the predicted time frame.
6. At 20 of the 24 sites, chemical data collected during remedial operation exhibited trends consistent with the presence of dense non-aqueous phase liquids (DNAPLs). However, even where substantial soil and water quality data were available, a separate immiscible phase was rarely sampled or observed. This is consistent with DNAPL behavior; i.e., they can move preferentially through very discrete pathways that easily may be missed even in thorough sampling schemes. DNAPLs were observed at sites where contaminant concentrations in groundwater were less than 15% of the respective solubilities.
7. The importance of treating groundwater remediation as an iterative process, requiring ongoing evaluation of system design, remediation time frames, and data collection needs, was recognized at all of the sites where remedial action was continuing.

The EPA is very careful in the report to state that none of the sites selected had their extraction systems optimized. Even though the contaminant concentrations seem to be stabilized at 17 of the sites, the EPA states: "The apparent stabilization of contaminant concentrations may be due to a number of factors not necessarily related to technical limitations of groundwater extraction. These include non-representative monitoring techniques, other contaminant sources not previously identified, inadequate extraction network design, and/or inefficient operation of the extraction network." In this report, the EPA has not given up on pump-and-treat as a remediation technology.

The National Research Council reviewed 77 cases for pump-and-treat and found that this method was able to achieve full cleanup at only eight sites. The committee came to the conclusion that pump-and-treat was ineffective at locations that contained significant amounts of solvents, precipitated metals, contaminants that have diffused into small pore spaces, or those that adhere strongly to soils (National Research Council, 1994).

The longest-running pump-and-treat system that Geraghty & Miller, Inc., has worked on is shown in Figure 18. Geraghty & Miller has been operating a pump-and-treat system for a client in the northeast since 1978. The system was originally installed to control and remediate a carbon tetrachloride contamination plume. There are several interesting features that can be depicted from Figure 18. First, Well 1 was the downstream well originally constructed to ensure that the carbon tetrachloride plume would not spread further. It has successfully performed this task and the plume has not spread from that well site. In 1982 a second well was installed upstream of Well 1 to help remediate the site faster. Well 2 successfully cut off all source of carbon tetrachloride to the aquifer entering Well 1. As can be seen in Figure 18, the concentration in Well 1 seems to go through two periods in which it stabilizes. The first period is prior to the installation of Well 2 when the concentration seemed to stabilize at about 20 µg/L. The second period is after the source of contamination is prevented from reaching Well 1, when the concentration seems to stabilize at about 5 µg/L. Even with the original source of carbon tetrachloride being removed from Well 1, the carbon tetrachloride concentration still stabilizes and is not completely cleaned out of that area of the aquifer. Well 2 also goes through stabilization at approximately 40 µg/L. As can be seen, the performance of this pump-and-treat system coincides with the EPA review of the 24 sites.

Other theoretical and laboratory studies have shown the same problems. Feenstra performed a series of laboratory tests on the influence of DNAPL on the performance of groundwater pump-and-treat systems (Feenstra, 1992). In the laboratory he had ground water flow directly through areas contaminated with DNAPL and had water flow parallel to pools of DNAPL. In his studies he found that DNAPLs could continue to act as a source of contamination for over 1000 pore volumes of water movement. His two main conclusions were, "(1) DNAPL zones represent continuing sources, and (2) pump and treat are unlikely to accelerate dissolution significantly."

While Feenstra and the EPA have concentrated on DNAPLs as the main source of continuing contamination at sites, the results shown by the Geraghty & Miller pump-and-treat and others show that geological conditions can also cause these stabilized concentrations in the pump-and-treat system. Goltz and Roberts conclude, "Relatively small zones of immobile water may result in extensive tailing, if sufficient sorption capacity exists in the zones." (Goltz, 1986).

Whatever the reason at a particular site, pump-and-treat systems have not been shown to be able to completely remove contaminants from an aquifer. None of these results should have been surprising if water would have been reviewed

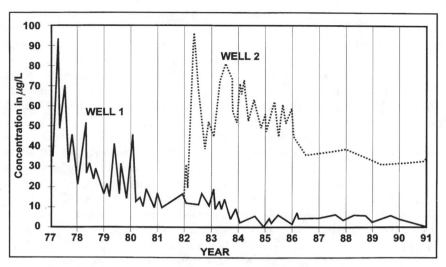

Figure 18   Results from a carbon tetrachloride pump and treat.

as the carrier in the pump-and-treat systems. As has been discussed in this chapter, the severe, natural limitations of water as a carrier would produce the results that we are now finding in full-scale pump-and-treat systems. No aquifer material is perfectly homogenous; therefore, the water cannot come into intimate contact with all portions of the aquifer. While people originally designed systems for advection to control the remediation of a plume, advection only controls when the plume is spreading. Because of the nature of the subsurface materials in an aquifer, diffusion controls when we are trying to use water as carrier to remediate a site. The advection process will allow us to remove large masses of contaminants, while the diffusion process will prevent us from cleaning the last residual of the contaminants from the aquifer.

This does not make pump-and-treat a worthless tool in remediation. Pump and treat is very effective, if designed correctly, in stopping the movement of plumes and for certain other specific functions, as in creating a cone of depression to help remove LNAPLs. Remember, the movement of plumes is an advection process, and pump-and-treat is an excellent advection controlling device. Pump and treat is also successful at removing large masses of contaminants from the aquifer. The failure of pump-and-treat to remove the last residual should not prevent the use of pump-and-treat in the areas in which it can be successfully applied. But, it must be remembered that the main limitation of pump-and-treat is that water is the carrier.

AIR AS THE CARRIER

The most important advances in the remediation field in the last five years have all been based upon using air as the carrier for removal of contaminants.

The two hottest "*in situ*" technologies being applied in the field today are soil vapor extraction (SVE; also known as vapor extraction system, VES), and air sparging. Neither of these techniques is, in fact, an *in situ* method. Both of these technologies rely on air movement to remove the contaminants from the ground and aquifer. This does not constitute an "in place" treatment; it is a simple change of carrier. SVE and air sparging use air as the carrier to remove contaminants from the ground and aquifer. Air provides several advantages over water but still has some of the weaknesses of water.

One of the important advantages in switching from water to air as the carrier is the number of pore volumes that can be processed through the soil or aquifer in a short period of time. Let us compare water and air based upon pore volumes in the same geological setting. The detailed calculations for the following example can be found in Nyer and Schafer's 1994 paper (see Nyer and Schafer, 1994).

Figure 19 shows the plume and capture zone boundary for a 10-gpm well located 25 ft from the leading edge of the plume. Analytic element modeling was used to determine the flushing rate by computing the travel time from the upgradient edge of the plume to the recovery well. Figure 20 shows a streamline approaching the recovery well and includes tick marks at 10-day intervals. The total travel time from the upgradient edge of the plume to the recovery well is 175 days based on an assumed porosity of 25%. Thus, a conservative estimate of pore volume exchange rate is 1 every 175 days.

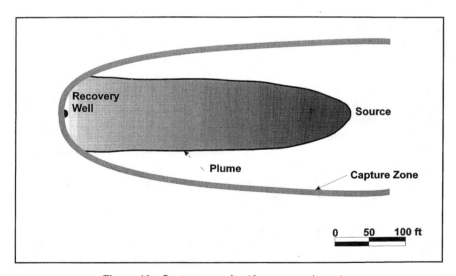

Figure 19 Capture zone for 10-gpm pumping rate.

During this time, however, clean water outside the plume, but within the capture zone, converges through the plume to the recovery well. This has the effect of increasing the number of pore volume flushings. To account for this, another way to estimate the number of pore volume exchanges is to simply compare the volume of water in the plume to the extraction rate of 10 gpm.

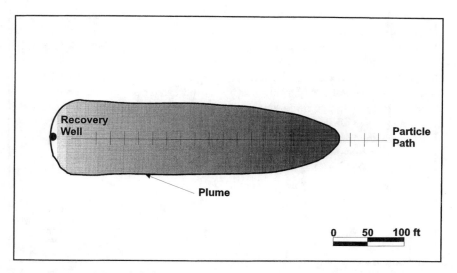

Figure 20 Ten-day time intervals along streamline show 175-day travel time from upgradient edge of plume to extraction well.

Approximating the plume as an 80- by 300-ft ellipse, the volume of water it contains is expressed as follows:

$$V = \pi \cdot \frac{300}{2} \cdot \frac{80}{2} \cdot 40 \cdot 0.25$$

$$= 188,495 \text{ ft}^3$$

Dividing by the flow rate of 1920 cfd (10 gpm) gives an average flushing time of 98 days. We will use this number is comparing pore volume exchanges.

Assume now that we have a similar area of soil contamination above the water table that will be cleaned up using vapor extraction wells. Assume further that the thickness of the vadose zone is 40 ft, the same as the assumed aquifer thickness in the previous example. Figure 21 shows a typical design we might use to vapor-extract contaminants using four wells running along the axis of the contaminated area. Often we select flow rates for the vapor extraction wells to produce from one to four pore volume exchanges per day. For this example, we will aim for two pore volume exchanges per day.

Using the same porosity as before, 25%, the air volume within the contaminated zone is 188,495 ft³. The total air flow rate required for two pore volume exchanges per day is

$$Q = \frac{2 \cdot 188,495}{1,440}$$

$$= 262 \text{ cfm}$$

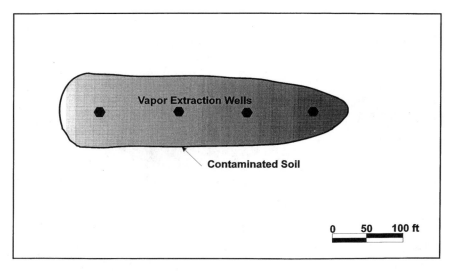

Figure 21   Locations of four vapor extraction wells.

This requires an average flow rate of 65.5 cfm per well. To see if this is a reasonable expectation, Nyer and Schafer used the Hantush leaky equation to estimate the drawdown (vacuum) associated with operating vapor extraction wells at this flow rate. Based on this analysis, and assuming the vapor extraction well efficiency will range between 20 to 80%, the calculated vacuum inside the wells would be expected to range between 15.1 in. of water (80% efficiency) and 60.3 in. of water (20% efficiency). There would be an additional vacuum component caused by interference by the adjacent wells, but this contribution is minor and may be ignored.

Commercial blowers are readily available to sustain the desired yield at either of the calculated vacuum values, so the projected design is a good one, assuming reasonable well efficiencies are obtained. The Hantush equation can also be used to demonstrate that adequate vacuum is achieved everywhere within the contaminated area.

Thus, the target of two pore volume exchanges per day can be realized under the assumed conditions. Comparing two air exchanges per day with one water exchange per 98 days, it is clear we have a 196-fold increase in the rate of pore volume exchanges with the vapor extraction system.

Many SVE and air sparging systems clean a site in 6 to 24 months, or at least, reach equilibrium in that time. We expect the pump-and-treat method to take between 5 and 20 years. As can be seen, pore volumes exchange are a significant part of the advantage. Of course, chemical properties can add to the advantage. Sims concluded, "... movement of volatile organic chemicals (VOCs) is generally 10,000 times faster in a gas phase than in a water phase..." (Sims, May 1990). The 10,000 factor comes from a combination of pore volume exchange rate, solubility versus volatility, and several other factors involved in the dynamic equilibrium of the chemicals in the ground. Hansen summarized the dynamic equilibrium of contaminant partitioning. (Hansen, Flavin, and Fam, 1994.) VOCs partition into four distinct yet interrelated phases: (1) NAPLs adsorbed to saturated or unsaturated soil particles, (2) free phase NAPLs, (3) soluble constituents dissolved in the groundwater, and (4) volatile constituents in the soil pore space of the vadose zone. All four phases are in equilibrium. Air or water as the carrier must interact with these phases in order to remove the mass from ground.

The air can come into direct contact with NAPL when it is in the free phase or adsorbed to the soil. The organic compounds can directly volatilize into the air from the NAPL. If the NAPL is located in one of the main flow paths of the air, then the organic will be directly removed from the ground. The rate of removal will be very high since the air will be continuously replaced and the driving gradient of the volatilization will always be at a maximum. Direct volatilization can occur even if the NAPL comes into contact with air that is not in a main flow path. The rate of volatilization will be less as the partial pressure of the organic increases and the driving gradient is reduced. The organic will have to diffuse to an area of high air flow in order to be removed from the ground. Most chlorinated organic compounds and many petroleum-based organic compounds are more volatile than they are soluble. This is a property that will allow the air to carry the organic compounds from the ground faster than water.

The NAPL can also be in direct contact with water when the NAPL is in the free phase or sorbed to the soil. This can be true in the vadose zone as well as the aquifer. The vadose zone is not saturated, but in many areas of the country it still contains a significant amount of water. Of course, desert areas have very little moisture in the vadose zone. The air would be in contact with the water, the organic compounds from the NAPL would dissolve into the water, and the compounds would then volatilize from the water to the air when air is the carrier. The rate of transfer to the air would be related to the volatility and the solubility of the compound. The Henry's Law constant would best relate the rate of transfer from the water to the air. The actual removal of the organic compound from the ground would be limited by both the rate of transfer from the NAPL to the water and the rate of transfer from the water to the air. The driving gradient for each transfer would be limited by how far away the organic compound was from a main air-flow path. Once again, compounds not in direct contact with a main air-flow path would have to diffuse to the air flow path. In the aquifer, the diffusion would occur in the water. In the vadose zone, the diffusion would occur in the air. Compounds that were dissolved in the ground water, but not from a NAPL source, would behave the same as compounds from a NAPL. The Henry's Law constant would control their transfer to the air, and diffusion would control their removal from non-flow areas.

The other function that air as a carrier can perform is to bring material into the ground. The best example of using air to carry material into the ground is oxygen transfer for biochemical reactions. As we discussed earlier in this chapter, bacteria use many of the organic contaminants as a source of food and energy. The reason that this reaction does not clean up aquifers by itself is that the rate of reaction is limited. The main rate-limiting factor is oxygen. Moisture content (in the vadose zone), nutrients, temperature, etc. will also limit the rate of bio-chemical reactions, (Chapter 3 will discuss biochemical reactions and *in situ* biological remediation). Atmospheric air is 20% oxygen. When the air is brought below ground with a SVE or sparging system, the oxygen can transfer to the water environment of the bacteria and supply the rate-limiting factor to the bacteria. As a comparison, water can only deliver 8 to 10 mg/L of oxygen. Once again, not only can we provide more pore volumes of air to carry the oxygen into the ground, the air also can carry significantly more oxygen with each pore volume. As we will discuss in Chapter 3, many *in situ* biological designs incor-porate air as the method to deliver oxygen to the contaminated zone.

LIMITATIONS

Air is still not the perfect carrier for remediation. There are several factors that limit the use of air. The same geological limitations occur with air as the carrier as occur with water as the carrier. The geology does not change just because the zone is unsaturated. There are still areas of preferred flow and relatively stagnant areas in SVE systems. The compounds still must diffuse over to an area of air movement to be removed from the ground. Diffusion still controls the end of the project.

Air sparging also has been shown to suffer some of the same limitations as pump-and-treat. Recent tests have proven that the air moves through channels in the aquifer. These channels do not change over the course of the project even if the system is turned off and then on again. The air has preferred paths that it travels. Any contaminants that do not come into direct contact with those channels have to diffuse to a channel in order to be removed from the aquifer by the air carrier. As with water movement, this effect creates the same flattening of the concentration curve after the mass removal phase. Diffusion still controls the process after initial mass removal and helps to determine whether the contami-nants reach the cleanup goal within a reasonable time frame. The air is able to remove more mass over a short period of time, but in the end, diffusion still controls the final concentration in the aquifer.

Air also has physical and chemical limitations as a carrier. Not all contam-inants can use air as a carrier. If a compound is not volatile, then air will have very limited use. High molecular weight organic compounds and metals will not be able to use air as a carrier. These types of compounds will be left in the ground as the air moves through the contamination zone. Even when compounds are volatile, air may have limitations. If the organic is also very soluble (i.e., ketones, alcohols, etc.), then air will not be able to remove the compound from the water. Organic compounds with low Henry's Law constants are not removed from water by air strippers in the above-mentioned ground treatment systems. Air also will not work below ground on these compounds when they dissolve in water.

There are also limitations when air is used as a carrier to deliver material to the ground. While oxygen is the main limiting factor for biological reactions in the ground, the other factors can also limit reaction rate. Nutrients, nitrogen, and phosphorous cannot be carried by air. There has been some limited work on using ammonia and nitrous oxide in the gas phase to deliver nitrogen through the air, but, in general, air is not used for nutrient delivery.

Even when oxygen is the component that needs to be delivered, air has limitations. We have discussed in detail that carriers do not flow past every particle of soil in the ground. The same holds true for using air as a delivery system. Once again, only diffusion is available to transfer the compounds into the low flow and stagnant areas. Oxygen must diffuse through the air or water in order to get to every microenvironment in which bacteria are degrading organic contaminants. Diffusion will limit the rate of oxygen delivery and the subsequent rate of biological degradation.

CONCLUSION

One principle that is consistent throughout all of the *in situ* technologies is the use of carriers for delivery and removal. Mass transfer and diffusion limitations will be a part of every project. The mass transfer portion of the project is usually at the beginning of the remediation. This part of the project can show a significant advantage by switching carriers from water to air. Enhancements to the carrier, i.e., temperature, surfactants, etc., will also show significant effect during the mass transfer portion of the project. However, we must consider the entire remediation project when analyzing or designing an *in situ* method of remediation. Diffusion effects must be taken into consideration, and we should address the question of reaching "clean" by the *in situ* method when we design it. We should not wait another 5 to 10 years before we decide if these new *in situ* methods will reach "clean".

This book will review each *in situ* technology using the "carrier" idea to describe how each technique works. This will allow an easier comparison between each technology. The basic comparison for each method will be: What is the carrier? How effective is mass removal with the method? What will limit the method from reaching "clean"?

We have spent years discovering the limitation of pump-and-treat. We must understand the replacement technologies better when we install them. We do not want to wait several more years before we realize that there are still significant limitations. This book is oriented toward understanding the limitations now so that we know when and how to install the new *in situ* technologies. We also need to develop a reasonable expectation of what these new technologies can accomplish.

REFERENCES

EPA. Evaluation of Ground-Water Extraction Remedies: Phase II. Volume 1 Summary Report, Publication 9355.4-05. Washington, D.C.: U.S. Environmental Protection Agency, Office of Emergency and Remedial Response, 1992a.

Feenstra, S. "Influence of DNAPL on Performance of Groundwater Pump-and-Treat Remedies." Presented at the National Ground Water Association Annual Meeting and Exposition, Las Vegas, Nevada, October 1992.

Goltz, M.N. and Roberts, Paul, V. *Interpreting Organic Solute Transport Data from a Field Experiment Using Physical Nonequilibrium Models,* Elsevier Science Publishers B.V., 1986.

Hansen, M.A., Flavin, M.D., and Fam, S.A. "Vapor Extraction/Vacuum-Enhanced Groundwater Recovery: A High-Vacuum Approach," Geraghty & Miller internal paper, 1994.

Keely, J.F. "Performance Evaluations of Pump-and-Treat Remediations." *EPA Ground Water Issue,* October 1989.

National Research Council, *Alternatives for Ground Water Cleanup,* Washington, DC.: National Academy Press, 1994.

Nyer, E.K. and Schafer, D.C., "There are No *In Situ* Methods," *Ground Water Monitoring and Remediation,* Fall 1994.

Perry, J.J. "Microbial Metabolism Of Cyclic Alkanes," in *Petroleum Microbiology,* R.M. Atlas, Ed., New York: Macmillan, 1984, pp. 61-98.

Roberts, P.V., Goltz, M.N., and Mackay, D.M. "A Natural Gradient Experiment on Solute Transport in a Sand Aquifer 3. Retardation Estimates and Mass Balances for Organic Solutes." *Water Resour. Res.,* vol. 22, no. 13, pp. 2047-2058, 1986.

Sims, R.C. "Soil Remediation Techniques at Uncontrolled Hazardous Waste Sites. A Critical Review." *J. Air Waste Manage. Assoc.,* vol. 40, no. 5, 1990.

Szecsody, J.E. and Bales, R.C., *J. Contam. Hydrogeo.,* vol. 4, pp. 181-203, 1989.

2 LIFE CYCLE DESIGN

Evan K. Nyer

The life cycle concept helps to focus the designer on the main strategies necessary to successfully remediate a site. The concept of Life Cycle design and its use on groundwater was first published in *Groundwater Treatment Technology* (Nyer, 1985). The simple basis for the life cycle concept is that groundwater remediations are unique, and the requirements for the project will change over the life of the project. One must design for the entire life of the project not just the conditions found at the beginning. Since 1985 we have continued to apply the concept of life cycle to groundwater treatment designs. However, over those years there have been three major interpretations of the life cycle. This chapter will review each of the main interpretations of the life cycle of a groundwater remediation and within the review show how design concepts have changed in groundwater remediation over the last decade.

There are two reasons that this book contains an entire chapter on the life cycle design concept. First, the life cycle, as originally described in 1985, was an early indicator that we would not be able to reach "clean" with pump-and-treat systems. Understanding the life cycle of a groundwater remediation will help us understand the limitations of pump-and-treat and the possible limitations of an *in situ* treatment method. Second, *in situ* treatment remediations will go through a life cycle. Once again, the conditions at the beginning of the project will not be the same as the conditions encountered during the middle of the project, and they will continue to change as the project progresses to the end. The design of the *in situ* remediation must encompass all of the conditions to be found during the remediation and not be solely based upon the initial conditions.

LIFE CYCLE DESIGN FOR PUMP-AND-TREAT SYSTEMS

In 1985, the main treatment technology was pump-and-treat. There was very little discussion on remediation methods. The main discussion was on the type

0-87371-995-6/96/$0.00+$.50
© 1996 by CRC Press, Inc.

of technology used to remove the organics and metals from the water withdrawn as the result of the pump-and-treat system. The main use of the life cycle curve was to provide a model that could be used to design the groundwater treatment system. There were three main lessons learned from using the original life cycle model: (1) concentration change over time, (2) capital costs are an important consideration because of the limited time each piece of equipment will be used, and (3) operator expenses are a significant part of the treatment costs.

Concentration Changes with Time

There are three patterns that contaminant concentrations follow over the life of the project. These patterns are summarized in Figure 1. First, there is the constant concentration exhibited by a leachate. If we do not remove the source of contamination, then the source will replace the contaminants as fast as they can be removed with the groundwater pumping system. Until the source of contamination is remediated, the concentration will remain the same. We normally think of "mine" leachate or "landfill" leachate. But, anytime there is a continuous source of contamination, we are dealing with a leachate. A non-aqueous phase liquid (NAPL) or a large clay lens impregnated with dissolved contaminants can also represent a source of contamination. As the water moves around the clay lens and/or the NAPL diffuses into the ground water, a well downstream of the source will show a continuous concentration of the contaminant.

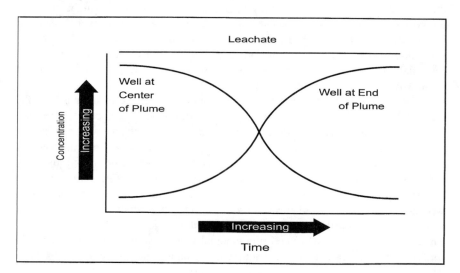

Figure 1 Time effect on concentration.

Several other examples of sources have developed as we have gained experience with remediation. Light non-aqueous phase liquids (LNAPLs) have been found to be a continual source in two major locations. First, they are sorbed in the vadose zone. Rain water (or other surface water) can cause a vertical migration of the contaminants into the ground water. Second, the "smear zone" can act as a source of continual contaminants. The smear zone is created by the change in groundwater levels, creating a subsequent change in the level of the LNAPL floating on top of the water. This allows the LNAPL to sorb to the soil over a wide vertical zone. When the water rises again, the sorbed LNAPL does not float out of the soil; it stays sorbed to the soil. This creates an area that has relatively low permeability to both water and air carriers, making it difficult to remediate. However, the smear zone is still in contact with both the vadose zone and the aquifer, and the organic compounds making up the LNAPL can diffuse into either area. Dense non-aqueous phase liquids (DNAPLs) can cause the same type of effect in the aquifer. As the DNAPL travels down through the aquifer, a portion is sorbed to the soil. While these areas may have a reduced permeability to water movement, water can still move through the affected zone, picking up contaminants. These zones can also act as a diffusion source of organic contaminants.

The second possible pattern arises when the contamination plume is being drawn toward the groundwater removal system. This mainly happens with municipal drinking water wells. In this situation, the concentration increases over time. The well is originally clean, but becomes more contaminated as the plume is drawn toward the well. It is very important to recognize when this situation will occur. Since the concentration will rise over time, the original treatment system must be over-designed to allow for increases in concentration. This will allow the treatment system to be designed for the entire life of the project. The curve that is shown in Figure 1 represents a large plume or a situation where the source of the contaminant has not been removed. Smaller plumes that have had their source of contamination controlled will increase and then decrease. The center of the plume will be drawn toward the low hydraulic head created by the large amount of pumping. But once the center of the plume is being pumped, then the concentration will start to decrease. This process can take a very long period of time, decades, or in some cases even centuries.

The final pattern is associated with remediation. In this case, the original source of contamination is removed. The pumping system is placed near the center of the plume. This should be the area of highest concentration, and the place where the water will bring the maximum amount of mass of contaminants to the withdrawal point for removal from the aquifer. As the pumping continues, the concentration of the contaminants decreases over time. The rate of decrease, very fast at the beginning of the project, slows and then finally stops decreasing or reaches an asymptote. We originally thought that this was the result of just retardation, natural chemical and biochemical reactions, and the dilution of the surrounding groundwater. As we discussed in Chapter 1, we now realize that the geology and micro flow patterns play an important role in the life cycle pattern of remediation. While the beginning part of the life cycle curve is concerned with the main body of the contamination, further along into the life cycle minor sources of contaminants control the shape of the curve. When we were designing our first groundwater treatment systems, we were only concerned with the beginning part of the life cycle curve. In fact, recognizing what occurred during the beginning of the remediation curve was a giant step toward proper design of groundwater treatment systems.

In the early 1980s, the main problem with groundwater treatment designs was that the concentration values used to determine the type of technology and treatment system size were overly conservative. It was common, at the time, to summarize all the concentrations found in the monitoring wells and to use the maximum concentration found in the highest concentration well as the initial concentration for the groundwater treatment system. This often led to the incorrect selection of technology. When the pumping wells were installed and the system finally turned on, the actual concentrations found at the influent to the treatment plant were significantly lower than the design concentrations.

Most treatment systems do not get more efficient as the influent concentration decreases. Metal removal and biological treatment systems can have a catastrophic failure if the influent concentration drops below a minimum level. The selection among other technologies can be based upon total pounds of contaminant that have to be removed. For example, one of the main costs of carbon adsorption is the replacement of the spent carbon. This is related to the total mass of contaminants that are removed. Carbon adsorption will be skewed as a high cost technology if the wrong concentration is employed in its evaluation. As a result, many treatment systems have failed to meet discharge standards or have not been economical on their first day of operation.

Even if the original design did work, this design approach produced treatment systems that were no longer effective after a very short period of time. In 1985, the life cycle concept was introduced to help designers realize that their treatment system designs would have to treat changing concentrations over the life of the equipment and the remedial program.

The concept of the concentration change over the life cycle of the project was promoted to show that the treatment design would have to be flexible on any groundwater treatment system installation. No matter what the type of contaminant or the geological setting, the life cycle curve of remediation was consistent. In 1985 we were mainly worried about the beginning portion of the life cycle curve because we were primarily interested in the design of groundwater treatment systems and the effect of the changing concentration on the actual design. We did not think too much about the later part of the curve. We were not sure if it was a period of very slow decrease in concentration and the life cycle curve would be a straight line if the time was put on a log scale, or if it was a true flattening of the curve and the concentrations had stopped decreasing. While several studies were already available to tell us that the curve was probably flat, they were mainly in the hydrogeological literature. The engineers and hydrogeologists were kept separate, at the time, and the design engineers were simply told to design a groundwater treatment system based upon the results of the remedial investigation. The first part of the curve was a major advance in the treatment design method; the concentration would decrease as the remediation progressed. The last part of the curve was a simple guess, and we did not realize its importance at the time.

Capital Costs

Another factor that we faced in the early 1980s was a lack of experience in designing capital equipment for groundwater remediations. The best sources of experience, at the time, were engineers who had designed wastewater treatment systems. Most of the first designers transferred from the wastewater area. This was an experience similar to that of many of the hydrogeologists during the same period, who were transferring from the oil field. One of the main problems with wastewater as a background for the groundwater field was the length of time that the project would last. Municipal systems would be designed to last up to 50 years; industrial systems were expected to last at least 20 years. Most of the equipment used in the field will have a 5- to 20-year life expectancy. Municipal systems switched from steel tanks to concrete tanks in order to extend the life of those unit operations. Pumps and other equipment with moving parts have a lower life expectancy, and tanks and reaction vessels have a longer life expectancy. The cost of equipment in wastewater treatment is figured over the life expectancy of the equipment. However, the cost of equipment for a groundwater cleanup must be based on the time used on the project with an upper limitation being the life expectancy of the equipment.

In Chapter 1, we have already discussed that the total time for the mass removal portion of a cleanup would probably be much less than the twenty years necessary for an industrial wastewater project. In the previous section of this chapter, we saw that even if the life of the project is 10 years, all of the equipment would probably not be needed for the entire time. As the concentration decreases, some of the equipment would have completed its function. The second part of the life cycle design switched our thinking from the length of time that the equipment would last to the length of time the project would need the equipment. The difference can be significant.

In 1985, we were mainly interested in equipment associated with groundwater pump-and-treat systems. The example prepared back then was based upon a biological treatment system. The lesson still holds true for today for the type of equipment that we apply to groundwater remediations. The example provided below was produced in 1985 but will show the same results as the example that we will provide in the section of this chapter, Life Cycle for *In Situ* Projects. We have updated the interest rate in the example below, and the daily costs produced will be slightly different from the original 1985 calculations. (Nyer, 1995).

Let us assume that the cost of equipment for a submerged, fixed-film, biological treatment system is $100,000. If we set the amount of time that we need the equipment and the interest rate that we have to pay for the equipment, then we can calculate the daily cost of the equipment. One formula for calculating costs would be:

$$C = \frac{Cap}{\left[1 - (1+i)^{-n}\right]/i}$$

where C is cost per time period n; Cap is capital cost ($100,000 in our example); i is the interest rate; and n is the period of time.

We will assume that the interest rate is 9%. If the equipment is used for 10 years, the daily cost is $43/day. If the equipment is only needed for 5 years, the daily cost is $70/day. At 2 years, the daily cost is $156/day, and at 1 year, the daily cost is $299/day. All of these figures assume that we have no use for the equipment after its usefulness is finished on this project. Figure 2 summarizes the daily cost of equipment when used for various periods of time.

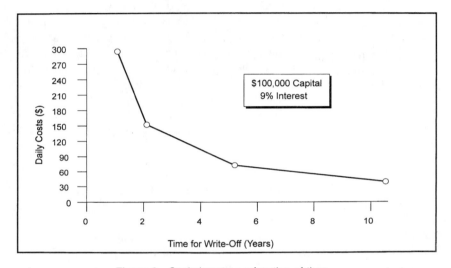

Figure 2   Capital cost as a function of time.

As can be seen, the cost of equipment gets significantly higher as the time of use decreases. The normal method of comparing the cost of treatment by different technologies is to base the comparison on cost of treatment per 1000 gallons of water treated. At a flow of 25,000 gal/day, the cost of treatment goes from $1.72 per 1000 gallons at 10 years to $11.96 per 1000 gallons at 1 year. Using the treatment equipment for 1 year will cost six times as much per gallon treated as using the same equipment for 10 years.

A great many groundwater cleanups will be completed in under 10 years, and many more will not use all of the equipment for the entire life of the project. This makes the cost of equipment over time another part of the life cycle design. The design engineer will have a problem on the shorter projects and on the longer projects in which a particular piece of equipment is only needed for a short period of time. An obvious solution to short-term use is to rent the equipment or to use it over several different projects. This would allow the equipment to be capitalized over 10 years even though it was only required for one year on a particular project.

Of course, any equipment that is to be used for more than one project will have to be transported from one site to the next. The equipment will have to be portable. For example, the design engineer needs a 15,000-gallon storage tank.

There is a choice of one tank 17 ft in diameter and 10 ft in height or two tanks 12 ft in diameter and 10 ft in height. If the equipment is to be used only a short period of time, the proper choice is the two 12-ft diameter tanks. The legal limit for a wide load on a truck is 12 ft. In general, to be transported by truck the treatment equipment should also be less than 10 ft in height and 60 ft in length. Rail transport can take slightly wider, higher, and longer units, but to be able to reach most of the U.S., shipment by truck should be assumed in the design.

Most of the equipment used today on groundwater pump-and-treat systems, and, in fact, on all remediation systems, is portable. First, most of the pump-and-treat systems are for very small flows. A 100-gpm unit is considered a medium-to high-flow system. Even for larger flow systems it is not difficult to make an air stripper portable. A 750-gpm packed-tower air stripper would be significantly less than 12 ft in diameter. Biological units have been designed in rectangular tanks able to fit on trucks. All carbon adsorption units are portable. Other equipment also has taken on the shapes and limitations necessary to make them portable. In 1985, portability was introduced as part of the life cycle design requirements for groundwater treatment systems; today we accept portability as part of the unique requirements of most groundwater remediation systems.

In the mid-1990s a new practice has started to become acceptable. Many small remediation projects, i.e., gasoline stations, are starting to use equipment that no longer has a long life expectancy. Small systems can cost more to move then they are worth. Plastics and other, less expensive materials are being used for construction. The life expectancy of the equipment in these systems is on the order of 5 years. The equipment is thrown away after it is used at the site.

Operator Expenses

One final area that has to be discussed under life cycle design is operator expenses. Any system that requires operator attention will cost more to operate than a system that does not require operators. All wastewater treatment systems should have operator expenses factored into their design. With groundwater treatment systems, this factor takes on added importance. The main reasons for this importance are (1) the relative size of a groundwater treatment system, (2) the remote locations of many sites, and (3) many remediations occurring at properties that are no longer active or have been sold to new owners. Once again, the engineer cannot just take a design developed for wastewater treatment systems and reduce its size for groundwater treatment. Most groundwater treatment systems will be very small in comparison to wastewater treatment systems, and most wastewater systems are associated with an active industrial plant. The operator costs, therefore, become more significant when we are dealing with groundwater treatment system designs.

In the 1980s we thought that most groundwater treatment systems would require regular operator attention. In the 1990s we are designing many systems to work without regular operators. These new system designs use various automatic analyzers and telemetry systems in order to inform a central office if the system needs attention. Remote locations and inactive sites have forced this change in approach for many remediations.

The significance of the cost of operator attention can still be shown by analyzing the relative cost of regular operator attention. Let us look at the biological treatment system example once again. Assume that a 15 HP blower is required for the system at $0.06/kwh. In addition, chemicals and miscellaneous costs are $3.00/day. At a 10-year life for the equipment, the daily costs would be

Equipment	$43.00
Power	$29.00
Chemicals	$ 3.00
Total	$75.00

Figure 3 summarizes the relative costs for each category. Without any operator attention, the equipment represents about 60% of the daily cost of operation. The power is about 35%, and the chemicals are about 5% of the daily costs. Figure 4 shows what happens to this relationship if one operator is required for one 8-h shift per day and is paid, including benefits, $10.00/h. Now 50% of the daily cost is represented by operator costs. Equipment drops down to 30%, power to 18%, and chemicals to 2%. At just one shift per day, the operator is now the main expense of the treatment system.

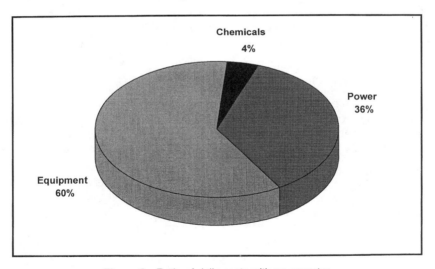

Figure 3   Ratio of daily costs with no operator.

If the treatment system requires full-time observation, the operator costs become even more important. Figure 5 shows the relative costs when an operator is required 24 h/day and paid $10.00/h. Now, the operator represents 75% of the cost of operation. Out of every $4.00 spent on the project, $3.00 would go to personnel.

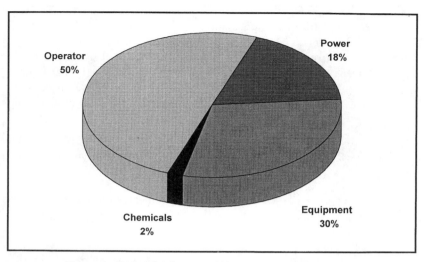

Figure 4 Ratio of daily costs with operator attention 8 h/day.

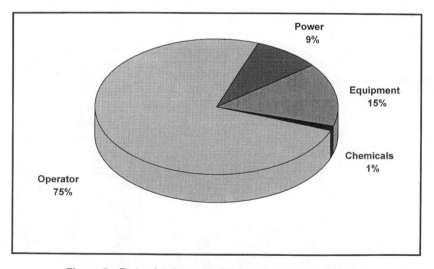

Figure 5 Ratio of daily costs with operator attention 24 h/day.

Daily costs for the project double if an operator is required for 8 h/day when compared to operating with no personnel. The costs triple at two shifts per day, and costs quadruple when around-the-clock attention is required. These costs are summarized in Figure 6. As can be seen from this data, the design engineer cannot ignore the effect of the operator on treatment system costs. Even when we extend the curve below the 8-h operator attention data point, we see that small amounts of operator attention can still add significant costs to the remediation. Designers should spend a significant amount of their effort on minimizing the operator time required for a particular design.

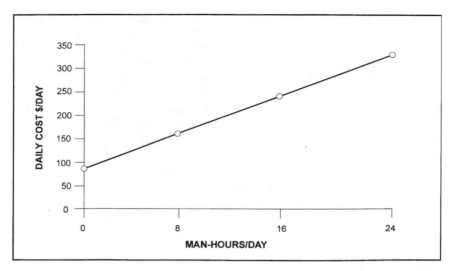

Figure 6 Daily cost of treatment with variable operator attention.

The effect of the operator does not decrease even as the size of the equipment increases significantly. Figure 7 represents the relative costs from a treatment system five times the size of the present example and requiring 24 h/day of operator attention. The operator still represents over one third the cost of treatment. Even as the total cost of the treatment system approaches $500,000, the design engineer must take special precaution to keep the required operator attention to a minimum.

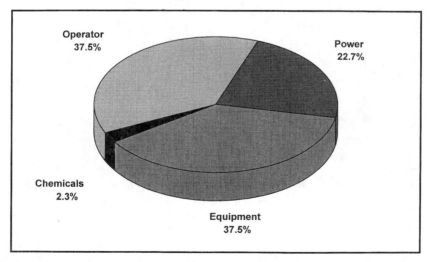

Figure 7 Ratio of daily costs for a $500,000 treatment system with operator attention 24 h/day.

In summary, there are three main factors that must be considered when determining a life cycle design for a groundwater treatment system. First, the concentration may change over time, the treatment design must meet the requirements at the beginning of the project, at the middle of the project, and at the end of the project. Second, because of the relatively short time that equipment is needed on groundwater projects, portable or inexpensive equipment should be considered. Finally, due to the relatively small size of groundwater equipment and the strong possibility of operation at an inactive site, manpower costs from operators become a significant, if not controlling, factor in equipment design (Nyer, 1989).

As you can see, all of the above discussions relate to groundwater treatment systems. It used to be a given that if the ground water was going to be pumped above ground and treated; it was assumed that this would clean the aquifer. The only question was what was the most cost-effective method of removing the contaminants once they were above ground. In the late 1980s, we realized that the pump-and-treat systems were not going to be able to complete the job. The next application of the life cycle curve was to try to understand the end of the project and to define it in a way that would allow the pumping system to be turned off.

USING LIFE CYCLE DESIGN TO DESCRIBE THE END OF THE PROJECT

In the late 1980s many pumping systems had been running for several years, and a couple of things started to become obvious. First, we were correct about the concentration decreasing as the project progressed. The concentration in the pumping wells and in the aquifer itself decreased as the pump-and-treat system operated. Second, the concentration curve flattened, and the concentration stopped decreasing after the systems had been running for extended periods of time. This caused two problems: (1) the pumping systems were no longer removing significant amounts of contaminants, but operational costs generally remained the same; and (2) the concentrations were not low enough to declare the aquifer clean and shut off the treatment system.

The life cycle curve was once again used to describe the change, or lack of change, in the concentration. In addition, the life cycle was used as a basis to develop a method to turn off the treatment equipment but still continue to make progress in the remediation of the aquifer. While the life cycle curve was still mainly concerned with above-ground pump-and-treat systems, our better understanding of *in situ* processes started to affect our interpretation of the curve.

Figure 8 shows the normal life cycle concentration of a remediation project. This is the same as the "well at the center of the plume" curve found in Figure 1. We have already discussed the beginning portion of the curve, we now will consider the bottom part of the curve. As the project progresses, the rate of removal decreases. As you can see in Figure 8, the curve tends to become parallel to the horizontal axis. Several factors that contribute to the flattening of this curve were discussed in the first chapter of the book. In this chapter, let us build a framework to help us understand how the life cycle curve can help us to reach the end of the project.

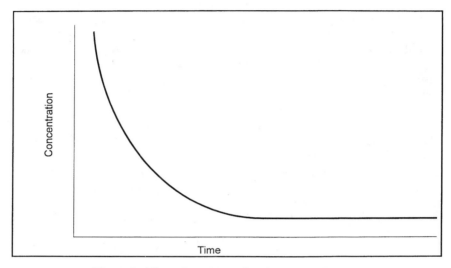

Figure 8 Life cycle concentration during remediation.

What is "Clean"?

The objective of most remediations is to clean the site. This raises the questions of what is "clean" and how do we define "clean"? The three main, conservative methods used to establish clean are: risk assessment, government regulations (federal, state, and/or local), and analytical detection limits. All of these methods have advantages and disadvantages.

Risk assessment can be uniquely designed for the specific site. Any unusual paths for human contact and any design method used to prevent human contact can be incorporated into the risk assessment. However, there are no official standard methods to develop the risk formulations. (At the time that this book was written, ASTM was just starting their standardization program.) Different basic assumptions will result in different specific numbers calculated for "clean". Some states will not allow risk assessment to be used because of the possible variable results from the same data. While risk assessment may be the only method that can consider local anomalies, the variable output may cause long discussions on the reality of the numbers.

Federal, state, and/or local regulations are another source of numbers that can be used to establish what is clean. The Clean Water Act has established maximum concentration levels (MCLs) for many compounds. Table 1 provides a list of current federal MCLs. These numbers are set and specific. Their bases were published and discussed before the final figures were made official. Over time, more compounds will be included on the list. The only problem for remediation sites is to establish the relationship between the soil and the groundwater. For example, all of the drinking water standards are based upon concentrations in water. The organics sorbed to the soil particles in the vadose zone and the aquifer will be released slowly into the groundwater. While drinking water standards have a strong technical basis, this method does not directly address the sorbed material.

Detection limits are the third method. For compounds that are highly toxic or in situations in which the numbers developed during a risk assessment are less than the detection limit of the compound, then the analytical detection limit can be used to establish what is clean. The main problem with this method is that the ability to detect a compound continuously improves. No one should accept "detection limits" as clean. Instead, the detection limits should be used as a basis for a specific concentration. Without a specific number, what is considered clean will change over the life of the project.

Any of these methods can be used to determine what is clean. However, if we look at the basis for each of these methods, we are basing these concentrations on contact with human beings. In other words, clean is when the groundwater or soil is safe for human consumption. The end of the remediation occurs when the site is safe for people.

A fourth method of determining clean has started to gain ground in the 1990s. Ecological risk assessments have been performed to determine the risk of harming some portion of the ecology when its environment is a receptor of the plume movement. For example, if the groundwater discharges into a surface stream, then the ecological risk assessment would determine if any of the fish or other fauna would be harmed by the contaminants. We will have to wait to see how pervasive ecological risk assessments become at remediation sites.

The problem with any of these definitions of "clean" is shown in Figure 9. As the site gets closer to "clean", the contaminant concentration reaches its asymptote. Figure 9 represents a worst-case scenario in which the site never reaches "clean". Even in cases where the site does reach "clean", it can take many years.

During the last years of the project, the treatment system suffers from diminishing returns. The treatment system continues to run, but the amount of material removed is minimal. The money is still being spent on the site, but the benefits are minimal for this financial outlay.

Retardation vs. Biochemical Activity

A second factor comes into consideration when we approach the asymptote of the life cycle concentration. Natural biochemical reactions may be occurring at the same rate as the natural release of the contaminants due to sorption and geological factors. This will be the diffusion-controlled portion of the project discussed in Chapter 1.

Table 1 Federal Primary and Secondary Drinking-Water Standards

Organic primary standards		Volatile organic primary standards		Microbiological primary standards		Physical primary standards		Radionuclides primary standards		Inorganic primary standards	
Contaminants	MCL (ppb)	Contaminants	MCL (ppb)	Contaminants	(per 100 mL)	Contaminants	MCL	Contaminants	MCL	Contaminants	MCL (ppb)
Alachor	2	Benzene	5	Bacteria	4	Turbidity	1 TU monthly avg. 5 TU avg. of 2 consecutive days	Gross alpha	15 pCi/L	Antimony	6
Atrazine	3	Carbon tetrachloride	5	Coliform	1			Manmade beta	4 milli-rem/yr	Arsenic	50
Benzo(a)pyrene	0.2	1,2-Dichloroethane	5					Radium 226 and 228	5 pCi/L	Asbestos	7.0 MFL
Carbofuran	40	1,1-Dichloroethylene	7							Barium	2000
Chlordane	2	*cis*-1,2-Dichloroethylene	70							Beryllium	4
2,4-D	70	*trans*-1,2-Dichloroethylene	100							Cadmium	5
Dalapon	200	Dichloromethane	5							Chromium	100
o-Dichlorobenzene	600	1,2-Dichloropropane	5							Cyanide	200
p-Dichlorobenzene	75	Ethylbenzene	700							Fluoride	4000
Dibromochloropropane	0.2	Monochlorobenzene	100							Mercury	2
Di(2-ethylhexyl)-adipate	400	Styrene	100							Nickel	100
Di(2-ethylhexyl)-phthalate	6	Tetrachloroethylene	5							Nitrate (as N)	10,000

Substance	Value	Substance	Value	Substance	Value
Dinoseb	7	Toluene	1000	Nitrite (as N)	1000
Diquat	20	1,1,1-Trichloro-ethane	200	Selenium	50
Endothall	100	1,1,2-Trichloro-ethane	5	Thallium	2
Endrin	2	Trichloro-ethylene	5		
Ethylene dibromide	0.05	Trihalomethanes (total)	100		
Glyphosate	700	Vinyl chloride	2		
Heptachlor	0.4	Xylenes (total)	10,000		
Heptachlor epoxide	0.2				
Hexachloro-benzene	1				
Hexachlorocyclo-pentadiene	50				
Lindane	0.2				
Methoxychlor	40				
Oxamyl (Vydate)	200				
Pentachloro-phenol	1				
Picloram	500				
Polychlorinated biphenyl	0.5				
Simazine	4				
2,3,7,8-TCDD (Dioxin)	0.00003				
Toxaphene	3				
2,4,5-TP (Silvex)	50				
1,2,4-Trichloro-benzene	70				

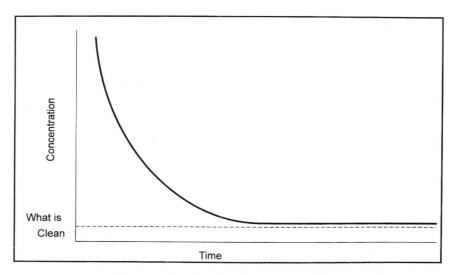

Figure 9 Reaching "clean" during remediation.

As we will discuss in Chapter 3, natural bacteria exist throughout the soil, vadose zone, and aquifer. If no toxic conditions exist, then the natural bacteria are already degrading the contaminants. Their rate of degradation is limited by the presence of a final electron acceptor (usually oxygen, but several final electron acceptors can be used by the bacteria; see Chapter 3 for a full discussion), nutrients, and the degradability of the organic compound. Oxygen is usually the main limiting factor. In fact, biodegradation models such as BIOPLUME II (Rafai et al., 1988) use oxygen concentration as the limiting factor in the degradation rate.

The natural rate of degradation is slow compared to an above ground reactor or to enhanced *in situ* biological reactions. The aquifer or vadose zone has a limited capability to replenish oxygen used by the bacteria. But, when the contaminant concentration reaches very low levels, then the aquifer and vadose zone can naturally supply what is needed. At the same time, all of the contaminants do not release to the water at the same rate. The contaminants are sorbed to the soil. Depending on the organic content of the soil and the chemical properties of the contaminant, the individual compounds will release to the water at a different rate. In addition, some of the contaminants will have to diffuse from their location to a place where they will be part of the main flow of the aquifer.

Other references (Rafai et al., 1988; Nyer, 1991) have shown that a plume of degradable organics would not move through an aerobic section of an aquifer. Data was from actual contaminated sites. Both projects showed that pumping would not improve the remediation. In fact, more compounds would be exposed to humans from a pumping system than by allowing the natural bacteria to degrade the compounds *in situ*. Both of these articles showed that natural degradation could be at the same rate as contaminant migration. We can project a complete natural elimination of the plume based upon the same data.

At some point, the rate at which the compounds can be removed from the aquifer by pumping will be equal to the natural biological degradation rate of the aquifer. At this point, we will not be able to speed the cleanup no matter how fast we pump the well. In fact, if the aquifer can naturally replenish the oxygen demand from the contaminants, then pumping will not increase the rate that we reach "clean" at all. Once we reach this point, all of the money that we are spending on pumping and treating this water is going to waste. The pumps could be turned off and all of the equipment removed, and we would still reach "clean" at the same time we would have by leaving the system running.

Active Management

In the late 1980s we tried to develop another point in the life cycle of the project that determines when we can turn off the treatment system. This is the point at which active management ceases and we can stop spending most of the money. "Clean" occurs when the contaminant concentrations have reached the point where they satisfy the regulator's cleanup criteria.

Figure 10 shows the life cycle curve with the clean line and the active management line. The active management line in Figure 10 represents when we can turn off the remediation equipment, or the point in the life cycle where pumping will no longer speed the cleanup of the site. We should stop spending money at this point. The clean line represents when we can use the water and the site for human consumption.

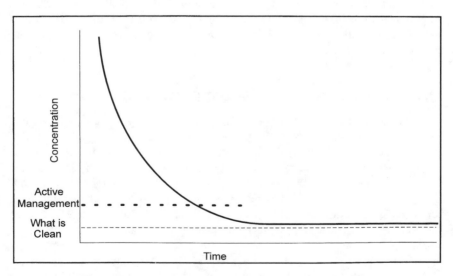

Figure 10   Active management end point during remediation.

This is very similar to situations that occurred in the 1970s when we cleaned up rivers and lakes. We installed wastewater treatment systems on municipal and industrial wastewater. This reduced the levels of contaminants entering the water body (lake or river) and allowed the river or lake to then clean itself. This took time, and we did not use the water body the day after we installed the treatment system. We waited until the water was clean or safe to use.

Rivers and lakes can remediate themselves faster than can groundwater. One reason for this is that oxygen transfers into these water bodies faster than it can into groundwater. But, groundwater *can* remediate itself. The problem is that the rate is so slow that we normally find the time frame unacceptable. However, if we remove most of the contaminants, then the aquifer can finish the job on its own. If the aquifer rate is the same as the rate from pumping and treating, then there is no reason to continue the pumping and treating. We have reached the point in the life cycle of the project when we simply must wait before we can use the water. Active management of the site should stop, and we should only monitor the site while we wait for "clean".

The only problem left is to determine when we reach the active management point of the life cycle of the project. For degradable organics, this point is basically a comparison of the normal fate and transport of chemicals to the biochemical reaction rate of those same chemicals. Some type of biomodeling will be required in conjunction with the solute transport model. Chlorinated hydrocarbons and metals will also undergo natural reactions that will contribute to their disappearance from the water. Modeling methods must include these rates of disappearance in order for the design to be able to determine when active management should end.

Life cycle design was still a tool of the pump-and-treat system in the late 1980s. But, as we started to recognize the limitations of water as a carrier, the importance of natural reactions occurring below ground, and the benefits of enhancing natural reactions, we started to incorporate *in situ* methods into our planning for the life cycle of the project.

LIFE CYCLE DESIGN FOR *IN SITU* TREATMENT METHODS

The current interpretation of the life cycle curve completes the incorporation of the *in situ* reactions into the planning of a remediation. Once we accept that most of the *in situ* technologies are based upon the use of water or air as the carrier, then the life cycle presentation of the change in concentration over the life of the remediation is accurate. Figure 1 represents the change in concentrations over the life of a vapor extraction system, sparging, or vacuum enhanced recovery. We will discuss each of these technologies in separate chapters later in the book. For this discussion, we just have to understand that all three technologies use air as the carrier.

As we discussed in Chapter 1, air has many advantages over water as the carrier in remediation. Air will be able to remove the organic compounds from the ground at a much faster rate than techniques that use water as the carrier. The simple advantage of air being able to move more pore volumes in the same amount of time would have the affect of shortening the amount of time to remove an organic compound. Assuming that the comparison is based upon two projects that had the same amount of organic originally in the ground, the concentration would decrease much faster with air as the carrier compared to water. The left side of the curve will have a steeper slope. (Of course, if we used pore volumes for the x axis, then the curves would be the same for water or air.) If the organic compound is volatile, then the curve would be even steeper with air as the carrier.

However, the right side of the curve would not automatically be affected with a switch from water to air as the carrier. The right side of the curve is controlled by diffusion and the geology of the site. Air has many of the same limitations as water. The material that is not in direct contact with the main flow path of the carrier must diffuse over to the main path in order to be removed from the ground. The right side of the curve would look the same for water or air. So while the slopes would change, the overall curve would still remain an accurate description of the change in concentration over the life of the project.

The life cycle curve has always been an important tool in the design of a remediation system. This is still true for *in situ* technologies that rely on one of the carriers. When we analyze the change in the shape of the life cycle curve between the two carriers and the various enhancements to the carriers, we realize that the left and right sides of the curve must be treated separately.

Figure 11 is the best way to think of the life cycle curve when designing a remediation. While the curve is the same, we have completely separated the left half and the right half of the life cycle line. The left half represents the mass removal portion of the remediation. The right half represents the reaching "clean" portion of the remediation.

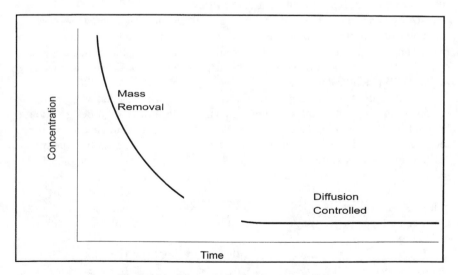

Figure 11 Life cycle for *in situ* remediations.

There are really two separate projects in any remediation. The first project is the removal of the maximum amount of mass of contaminants. The *in situ* technologies that rely on air as the carrier have been shown to have a significant advantage in this type of project. (Sims reported a factor of 10,000.) Most of the published information that we are currently receiving on *in situ* processes are really analyses of the mass removal capability of these techniques. The second project is reaching the mandated concentration required to declare the site "clean".

Pump and treat never failed as a technology for the first type of project. Water was capable of removing a significant portion of the mass of contaminants from the aquifer. Air may cost less and remove the compounds faster because of its advantages, but that does not mean that water failed. The Environmental Protection Agency (EPA) and the National Research Council (NRC) reviews were based upon the reaching-"clean" part of the remediation. When they stated that pump-and-treat was not capable of remediating an aquifer, they really meant that pump-and-treat failed to reach the concentration that would allow the aquifer to be designated as "clean". Pump and treat failed in the reaching-"clean" portion. (Both reports made it very clear that they felt pump-and-treat had been misapplied and that it was a successful technique for controlling plume movement.)

The real problem with pump-and-treat systems is that the designers expected that the systems would be able to clean the aquifer, and they told everyone that. They did not include the geology and diffusion limitations of using a carrier when they designed their remediation. Because of the slow pace of pore volume exchange with water as the carrier, the results did not show up for many years. This delay allowed thousands of remediation systems to be installed based upon the assumption that pump-and-treat could reach "clean".

Air as a carrier has the same geology and diffusion limitations. However, we are now installing the *in situ* technologies based upon air and expecting them to reach clean. We are taking the results from the first part of the remediation and declaring that since the compounds are being removed much faster then obviously the site will reach "clean".

This is why it is very important to separate the two projects. Mass removal is completely separate from reaching "clean", and a remediation design should be analyzed as two separate projects. What will the technique do for mass removal? What will the technique do for reaching "clean"? Separating the remediation into two separate projects does two major things for the designer. First, we are able to predict at the beginning of the project what will happen over the entire life of the remediation. Second, we are able to realize that different technologies may have to be applied to the site in order to complete the remediation. One technology may have to be used for the mass removal, and a different technology used to reach "clean".

Figure 11 is the representation of the life cycle of the remediation that we should be using in the 1990s. It is still important to use the life cycle curve when developing a strategy for a remediation. Figure 11 incorporates our latest understanding of the different factors that affect the entire remediation project.

DETERMINING THE TIME REQUIRED TO COMPLETE A LIFE CYCLE IN GROUNDWATER REMEDIATION

The last two questions that must be answered to complete the life cycle design are (1) when do we switch from mass removal to reaching "clean"? and (2) When is the project finally over? Both are very difficult questions. Too many people want to accept the answer, "You'll know it when you see it," and just start the project. Good design requires that the designer have some idea about when these two events will happen and at least a method of measuring when the events do occur.

The first question is easy to define on paper, but difficult to define in terms of a remediation project. The main requirements for the mass-removal portion of the remediation is to remove enough mass from the ground so that the method applied during the reaching-"clean" portion of the remediation can function. For example, if we were reviewing a compound that degrades aerobically and using natural attenuation to reach "clean", then oxygen demand becomes the key to the first portion of the project. We must remove enough organics during mass removal so that the oxygen demand below ground has been reduced. The level of reduction can be defined at the point at which the reaching-"clean" portion of the project can function without mechanical assistance to the natural oxygen transfer. The organics remaining in the ground will be degraded by the natural bacteria. The rate of degradation will be limited by the rate of natural oxygen transfer through the soil to the ground and aquifer. If the rate of natural oxygen transfer is sufficient to maintain an aerobic environment, then we have reached the point where the second portion of the project will work as natural attenuation. If we have left too much mass in the ground, this will create too much oxygen demand from the bacteria degrading all of that mass, and we will have to supply oxygen by mechanical means in order to maintain an aerobic environment. It is also easy to measure this example. We can simply measure the oxygen in the aquifer and the vadose zone and make sure it is above a certain minimal level that maintains an aerobic environment. If the oxygen is maintained at those levels when we turn off the equipment at the end of the mass-removal portion of the project, then we have successfully finished that project and we can move onto the second portion of the remediation.

While this example seems very simple, this concept is still relatively new. One of the authors was working on a project that tried to use this concept as the method to design and control the remediation at a site. While the overall life cycle concept of two separate projects was accepted, the problem occurred when we tried to define how to switch from mass removal to the reaching-"clean" portion of the remediation. The technical support for the regulators was of the opinion that mass removal should be continued until it no longer was removing a significant amount of material. The author tried to use the above description as a method from switching from mass removal to reaching clean.

The regulators suggested two ways to define the point at which mass removal was no longer removing a significant amount of material. The first was a simple pounds-per-day or concentration-per-well evaluation of what was being removed with the vapor extraction system (VES) system. The second method was to do a statistical analysis of the concentration coming out of the VES well and run the system until an asymptote was reached. The first method can suffer from background organic concentrations that show up in the VES system. These background concentrations, especially near gasoline stations, can make set concentrations or mass removal per day values very difficult to delineate. If the concentration is set too low, this can add years to the mass removal operation of the project without any significant decrease in mass in the ground.

The second method suffers from analytical costs. Developing enough data over a large site with multiple wells is always difficult. The number of data points and sampling events that would be required to statistically show that the data had reached an asymptote over a period of time can be very expensive. On this one project, the analytical costs of determining the asymptote were more expensive than the actual operations of mass removal.

This project pointed out that the first step in determining the life cycle design of the mass-removal portion of the project is to determine the objectives of that portion of the project. If the stated objective is to remove as much as possible, then the two methods suggested by the regulatory agency (minimum mass per day or asymptote) are methods that can be used to determine the end of the project. However, if the objective of mass removal is to establish a favorable subsurface environment so that the reaching-"clean" portion of the project can be achieved naturally, then these two methods may extend the operation of the mass removal project unnecessarily.

Simple monitoring of the system may show that the natural attenuation of even nondegradable compounds is sufficient after most of the mass has been removed. One operation method that may satisfy the regulators is to not remove the mass removal equipment once the end of that project is thought to have been reached. The equipment can simply be turned off and the vadose zone and the groundwater monitored for the next several quarters. If the monitoring data shows that enough mass has not yet been removed, then the mass removal equipment can simply be turned back on. This method allows flexibility with the design. Because of all the variables associated with the organics and the geology, it is difficult to make exact predictions of when the project should be changed from one portion to the next portion of the remediation. The simple testing of the actual results while the equipment is still in place is probably the best method, at this time, of determining when to switch from one project to the other. Hopefully, with time and experience we will become much better at knowing when to switch from one project to the next. Of course, all parties will first have to agree on the original objective of each portion of the project.

The second question is much more difficult to answer: When will the project finally be over and the site declared "clean"? There are two problems with declaring a site "clean". First is point of compliance and second is understanding and modeling the natural reactions of the compounds in the vadose zone and aquifer. Point of compliance is extremely important when trying to determine if the site has reached "clean". Monitoring wells represent the concentration of the aquifer or vadose zone in the immediate area of that monitoring well. As we will discuss in Chapter 3, there are several biochemical reactions that occur underground. If these reactions are fast enough, or at least as fast as the diffusion rate of the compounds, then material located in one portion of the aquifer may not reach the monitoring wells. Is this site still contaminated? Other work has fully explored this question (Nyer and Senz, 1995). The reader should look to these other publications for full discussion on this subject. For this chapter, let's just say that the point of compliance is extremely important as far as determining whether the site is determined to be clean.

The second portion of the question becomes when will a nondegradable compound reach "clean". From the work that the authors and others have done, especially if NAPLs are present, some sites may never reach "clean". This is very true for the center of the site where the contamination originally occurred. The term "never" refers to any site that takes over 50 to 100 years to remediate. Under these conditions the project really turns into a removal of mass and control of the site. The answer to the second question, then, is relatively black and white. Does the site have a chance to reach "clean", or does the nature of the contaminants and the aquifer force us to believe that the compounds will be slowly released to the aquifer over a very extended time. If the compounds are of a nature that the aquifer can reach clean, then the second part of the project is usually set up for natural attenuation and monitoring. Bio-modeling may be able to show, or at least provide a range of, the time it will take to finally reach the required concentration to declare the site clean. Monitoring is the only way to confirm that the actual conditions have been reached. When it is decided that the compounds cannot be removed successfully to the levels required, then the project switches over to a control project. The design objective switches from one of complete removal to one of complete control. One must design the system so that no compounds get off of the site. Pump and treat is a very good technology for control in an aquifer. New technologies, i.e., reactive walls, are coming along that may be more cost effective for this control. While the vadose zone and aquifer are both contaminated, the decision for removal or control will have to be separate for each zone.

Even the *in situ* technologies cannot escape the life cycle of a remediation. The life cycle concepts have taught us over the years how to apply technologies and design remediation. It is important to use Figure 11 when trying to design an *in situ* project. Separate the project into two major portions, mass removal and reaching "clean". Using this technique, the designer can incorporate the entire project into the original concept. The designer must decide in the beginning of the project what the objectives are and what can be accomplished at this particular site. The life cycle concept helps to focus the designer on the main strategies necessary to successfully remediate a site.

REFERENCES

Nyer, E.K. "The Effect of Time on Treatment Economics." *Ground Water Monitor. Rev.,* Spring 1989.

Nyer, E.K. "Biochemical Effects on Contaminant Fate and Transport." *Ground Water Monitor. Rev.,* Spring 1991.

Nyer, E.K. and Senz, C.D. "Is This Site Contaminated?" *Ground Water Monitor. Remed.,* Winter 1995.

Rafai, H. et. al. "Biodegradation Modeling at Aviation Fuel Spill Site." *J. Environ. Eng.,* vol. 114, no. 5, 1988.

Sims, R.C. "Soil Remediation Techniques at Uncontrolled Hazardous Waste Sites. A Critical Review." *J. Air Waste Manage. Assoc.,* vol. 40, no. 5, 1990.

3

IN SITU BIOREMEDIATION

Evan K. Nyer, Tom L. Crossman, and Gary Boettcher

INTRODUCTION

In situ bioremediation is a true *in situ* technology. In this process, biochemical reactions destroy the organic compounds. The actual reactions occur below ground, making this one of the only "in place" or *in situ* technologies used today. This is the main reason why *in situ* bioremediation is the first treatment technology that will be discussed in this book. The second reason is that many of the other *in situ* technologies rely on biochemical reactions to treat chemical constituents. The intent here is to provide a full understanding of bioremediation before describing the other technologies.

Treatment of organic compounds by biochemical reactions has been around for many years. The Clean Water Act forced many municipal and industrial facilities to use biological reactors to remove the organics from their effluents. Almost all of the organic compounds that we find below ground today have been or are being treated in above-ground biological reactors somewhere in the country. Based upon this experience, biochemical reactions should be a main part of cleaning up contaminated unsaturated zones and aquifers.

Over the years, this assumption has led to many applications of *in situ* bioremediation and to many failures. While there were a large number of people who understood the biochemical reactions, very few had a complete understanding of the environment in which the biochemical reactions were occurring. As we have discussed in Chapter 1, geology controls below ground, not the technology. Too many designers neglected geology as part of their bioremediation designs, and bioremediation obtained a mixed reputation.

During the 1990s several groups of researchers and designers from the Environmental Protection Agency (EPA), Department of Defense (DOD), large industrial companies, and universities combined biological and geological expertise and designed successful *in situ* bioremediations. Even with this success, a problem

0-87371-995-6/96/$0.00+$.50
© 1996 by CRC Press, Inc.

has occurred that still has the ability to taint the reputation of bioremediation: it is very difficult to "prove" that biochemical reactions were responsible for the destruction or disappearance of the organic compound.

Bacteria are ubiquitous. Researchers have found microorganisms throughout the subsurface. If there are any natural organics present in an unsaturated zone or an aquifer, a wide variety of microorganisms are also present. Even in areas that have low natural organic material, some amount of bacteria are usually present. These bacteria are found throughout the subsurface and can use organic contaminants as a source of food and energy. A general way to look at this is, "You don't bring remediation to a site; you only enhance the natural reactions that are already occurring." The challenge, then, becomes one of "proving" that the disappearance of the organic compound was due to biological activity when these reactions are part of the natural environment.

Even with these difficulties, *in situ* bioremediation has become a very important tool for the remediation of contaminated sites. The purpose of this chapter is to provide the reader with sufficient knowledge to be able to understand and apply biochemical reactions to the remediation of a site.

This chapter is divided into three main topics. We will start with the basic biochemical reactions. The biogeochemistry must be understood before we can interpret the site data and understand what is required to enhance the natural reactions. Second, we will review natural remediation, or "intrinsic" bioremediation. Many times the natural rate of bacterial destruction is sufficient to control and remediate a site. Finally, the chapter will review enhanced bioremediation. Enhanced bioremediation is necessary when the natural rate of biochemical reactions are not sufficient. The goal is to determine the rate-limiting factor and to design a mechanism to deliver the factor to the environment.

BIOCHEMICAL REACTIONS

The first area that we must understand is the biochemical reactions. The contaminants are not destroyed by magic microorganisms. We first have to understand these microorganisms, their reactions, and the environmental factors that can affect the microorganisms and/or their enzymes and reactions.

Microorganisms

Free-living microorganisms that exist on earth include bacteria, fungi, algae, protozoa, and metazoa. Viruses are also prevalent in the environment; however, these particles can only exist as parasites in living cells of other organisms and will not be discussed in this text. Microorganisms have a variety of characteristics that allow survival and distribution throughout the environment. They can be divided into two main groups. The eucaryotic cell is the unit of structure that exists in plants, metazoa animals, fungi, algae, and protozoa. The less complex procaryotic cell includes bacteria and cyanobacteria. Even though the protozoa and metazoa are important organisms that affect soil and water biology and chemistry, they do not perform important degradative roles. Therefore, this chapter will concentrate on bacteria and fungi.

Bacteria are by far the most prevalent and diverse organisms on earth. There are over 200 genera in the bacterial kingdom (*Bergey's Manual of Determinative Bacteriology,* 1974) These organisms lack nuclear membranes and do not contain internal compartmentalization by unit membrane systems. Bacteria range in size from approximately 0.5 μm to seldom greater than 5 μm in diameter. The cellular shape can be spherical, rod-shaped, filamentous, spiral, or helical. Reproduction is by binary fission; however, genetic material also can be exchanged between bacteria.

The fungi, which include molds, mildew, rusts, smuts, yeasts, mushrooms, and puffballs, constitute a diverse group of organisms living in fresh water and marine water but predominantly in soil or on dead plant material. Fungi are responsible for mineralizing organic carbon and decomposing woody material (cellulose and lignin). Reproduction occurs by sexual and asexual spores or by budding (yeasts).

Distribution and Occurrence of Microorganisms in the Environment

Due to their natural functions, microorganisms are found throughout the environment. Habitats that are suitable for higher plants and animals to survive will permit microorganisms to flourish. Even habitats that are adverse to higher life forms can support a diverse microorganism population. Soil, ground water, surface water, and air can support or transport microorganisms. For example, there are several genera of natural bacteria and fungi in soil and water capable of hydrocarbon degradation. The predominant bacteria genera (Dragun, 1988a; Kobayashi and Rittman, 1982) include *Pseudomonas, Bacillus, Arthrobacter, Alcaligenes, Corynebacterium, Flavorbacterium, Achromobacter, Micrococcus, Nocardia,* and *Mycobacterium.* The predominant fungal genera (Bumpus et al., 1985) include *Trichoderma, Penicillium, Asperigillus, Mortierella,* and *Phanerochaete.* Table 1 (Dragun, 1988) shows the microorganism population distribution in soil and ground water and demonstrates the variability and population sizes that can exist.

Soil

Bacteria outnumber the other organisms found in a typical soil. These organisms rapidly reproduce and constitute the majority of biomass in soil. Typically, microorganisms decrease with depth in the soil profile, as does organic matter. The population density does not continue to decrease to extinction with increasing depth, nor does it necessarily reach a constant declining density. Fluctuations in density commonly occur at lower horizons. In alluvial soils, populations fluctuate with textural changes; organisms are more numerous in silty or silty clay than in intervening sandy or coarse sandy horizons. In soil profiles above a perched water table, organisms are more numerous in the zone immediately above the water table than in higher zones (Paul and Clark, 1989). Most fungal species prefer the

Table 1   Microorganism Population Distribution in Soil and Ground Water

Organism	Population size	
	Typical	Extreme
Surface Soil (cells/gram soil)		
Bacteria	0.1 – 1 billion	> 10 billion
Actinomycetes	10 – 100 million	100 million
Fungi	0.1 – 1 million	20 million
Algae	10,000 – 100,000	3 million
Subsoil (cells/gram soil)		
Bacteria	1000 – 10,000,000	200 million
Ground water (cells per mL)		
Bacteria	100 - 200,000	1 million

upper soil profile. The rhizosphere (root zone) contains the greatest variety and number of microorganisms.

Ground Water

Microbial life occurs in aquifers. Bacteria exist in shallow-to-deep subsurface regions, but the origin of these organisms is unknown. They could have been deposited with sediments millions of years ago, or they may have migrated recently into the formations from surface soil. Bacteria tend not to travel long distances in fine soils but can travel long distances in coarse or fractured formations. These formations are susceptible to contamination by surface water and may carry pathogenic organisms into aquifer systems from sewage discharge, landfill leachate, and polluted water (Bouwer, 1978).

In all of these ecosystems, multiple types of bacteria exist. In fact, all degradation processes require multiple microorganisms working in concert. Also, more than one type of bacteria or fungi can perform the same degradation function. When investigating microorganisms for degradation of specific organic compounds, it is more important to demonstrate the ability to degrade a compound than to locate a specific bacteria or fungi.

Biochemical Reactions of Microorganisms

Microorganisms degrade organic compounds to obtain energy that is conserved in the C–C bonds of the compounds. The organics are converted to simpler organic compounds and, ultimately, carbon dioxide or methane and water. The microbes will also use part of the compounds as building blocks for new microbial cells.

Biodegradation is the process in which indigenous microorganisms convert or degrade natural and manmade organic compounds. Carbon sources not produced by any natural enzymatic process or having unnatural structural features are considered xenobiotic.

The main goal of a bioremediation design is the destruction of organic contaminants. However, we must remember that this is not the main goal of the bacteria and the fungi. The main function of bacteria and fungi is the degradation of natural organic material. This, in turn, is part of the natural carbon cycle of the earth (Figure 1).

As can be seen in Figure 1, the microorganisms perform a small part of the overall carbon cycle. When we discuss hazardous waste destruction, we are referring to a small part of microorganism activity: we are simply recycling manmade carbon compounds back into the existing carbon cycle.

Enzymes

Biodegradation of organic compounds (and maintenance of life-sustaining processes) are reliant upon enzymes. The best way to understand enzyme reactions is to think of them as a lock and key. Figure 2 shows how only an enzyme with the right shape (and chemistry) can function as a key for the organic reactions. The lock and key in the real world are three dimensional. The fit between the two has to be precise.

Organic compounds in the environment that are degradable align favorably with the active site of specific enzymes. Compounds that do not align favorably or compounds that do not bind with the active site of the enzyme will not be affected by the microorganism. Degradation of these compounds require that the microorganism population adapt in response to the environment by synthesizing enzymes capable of catalyzing degradation of these compounds.

Oxidation/Reduction

Organisms generally derive energy from oxidation-reduction (redox) reactions (catabolism). An enzyme-mediated oxidation/reduction reaction is the transfer of electrons from electron donors to acceptors. Energy is derived from these reactions when the energy source (electron donor) is oxidized, transferring electrons to an acceptor and releasing energy conserved in the chemical bond. Once the electron donor has been completely oxidized, the compound is no longer a source of energy. Bioremediation processes where microorganisms are exploited to degrade xenobiotic compounds are identical to natural degradative processes requiring enzymes. The energy released from these compounds is used by the organism to maintain life-sustaining processes.

There are three mechanisms used by microorganisms to produce energy. *Aerobic* respiration processes require oxygen; *anaerobic* processes rely on nitrate, iron, sulfate, or carbonate in the absence of oxygen to complete organic compound oxidation; and *fermentative* processes rely on organic compounds as electron donors and acceptors,. Microorganisms that require molecular oxygen are termed "obligately aerobic". These organisms cannot survive without oxygen. Within this group, microorganisms that survive on reduced oxygen concentrations are termed "microaerophilic". Although oxygen is the most common electron acceptor used

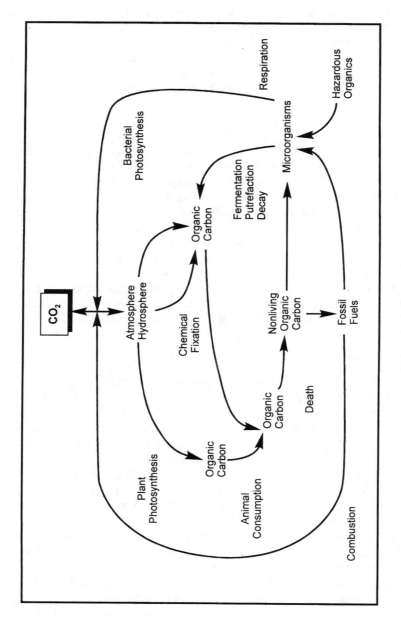

Figure 1 Carbon cycle.

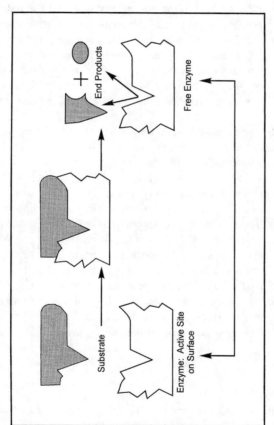

Figure 2 Enzymes represented as lock and key.

during respiration processes, nitrate is probably the most common alternate electron acceptor which is converted to more reduced forms of nitrogen. This process is called "denitrification" and most bacteria that reduce nitrate are called "facultative anaerobes" since they will transfer electrons to oxygen if it is present or to nitrate when oxygen is absent. The redox potential for the reduction of nitrate to nitrite is lower than the redox potential for the reduction of oxygen. Therefore, growth on nitrate is less efficient than growth on oxygen, and in most organisms nitrate reduction is strongly inhibited by oxygen. Sulfate is used by some bacteria as electron acceptors to produce hydrogen sulfide; however, these bacteria usually are strict anaerobes and are killed in the presence of oxygen. The redox potential for the reduction of sulfate is even lower than that for nitrate reduction, and efficiency of growth is also lower. Methanogenic bacteria (methane producers) are a diverse group of extremely oxygen-sensitive bacteria that reduce carbon dioxide to methane. Fermentation is the process of oxidizing some organic compounds in the absence of an added electron acceptor. Under fermentation processes, the organic is partially oxidized and only a small amount of energy is released.

Generally, an oxygen atmosphere in soil of less than 1% will change metabolism from aerobic to anaerobic (Paul and Clark, 1989). In aqueous environments, oxygen concentration less than approximately 1.0 mg/L can cause a switch of metabolism from aerobic to anaerobic (Tabak, 1981). Microaerophobic bacteria maintain aerobic reactions at reduced oxygen levels.

Recent work has shown that anaerobic reactions are an important part of natural degradation. Table 2 shows the aerobic and anaerobic pathways of benzene, toluene, ethyl benzene, and xylene (BTEX) biodegradation. When BTEX or other biodegradable organics are released, natural bacteria acclimate to produce requisite enzymes in order to liberate the energy conserved in the C–C bonds. These reactions require final electron acceptors as shown in Table 2. The bacteria uses oxygen until it is depleted and then switch to another electron acceptor. Table 3 shows the order in which the various electron acceptors are used by the microorganisms. The order is based upon the energy produced by the reactions (redox potential).

Inorganic Nutrients

Molecular composition of bacterial cells is fairly constant and indicates the requirements for growth. Water constitutes 80 to 90% of cellular weight and is always a major nutrient. The solid portion of the cell is made of carbon, oxygen, nitrogen, hydrogen, phosphorus, sulfur, and trace elements. The approximate elementary composition is shown in Table 4.

As can be seen from Table 4, the largest component of bacteria is carbon. The organics that we wish to destroy can provide this element. After carbon, oxygen is the highest percentage of the cell. When oxygen requirements of new cells is added to the required oxygen as an electron acceptor, large amounts of oxygen may be utilized in biological degradation.

Table 2   Aerobic and Anaerobic Pathways of BTEX Biodegradation

Pathway	Ref.
Aerobic respiration	
$(BTEX) + 9\,O_2 \rightarrow \quad 7CO_2 + 4H_2O$	(Rittmann et al., 1993)
Anaerobic respiration	
Nitrate reduction	
$5\,(BTEX) + 36\,NO_3^{-1} \rightarrow \quad 18\,N_2 + 35\,CO_2 + 38\,H_2O$	(Barbaro et al., 1991)
Manganese reduction	
$BTEX + 18\,MnO_2 + 29H^+ \rightarrow \quad 18\,Mn^{+2} + 7\,CO_2 + 22\,H_2O$	(Baedecker et al. 1993)
Iron reduction	
$(BTEX) + 36\,Fe^{+3} + 21\,H_2O \rightarrow \quad 36\,Fe^{+2} + 7\,CO_2 + 7\,H_2O$	(Loveley et al., 1989)
Sulfate reduction	
$8\,(BTEX) + 35\,SO_4^{-2} \rightarrow \quad 35\,S^{-2} + 56\,CO_2 + 28\,H_2O$	(Wilson et al., 1994)
Methanogenesis	
$(BTEX) + 42\,H_2O \rightarrow \quad 35\,CH_4 + 21\,CO_2$	(Wilson et al., 1994)

Table 3   Standard Reduction Potential at 25°C and pH 7 for Some Redox Couples that are Important Electron Acceptors in Microbial Respiration

Oxidized species	Reduced species	E° (volts)
$O_2 + 4H^+ + 4e^-$	$= 2H_2O$	+0.92
$2NO_3^- + 12H^+ + 10e^-$	$= N_2 + 6H_2O$	+0.74
$MnO_2(s) + HCO_3^- + 3H^+ + 2e^-$	$= MnCO_3(s) + 2H_2O$	+0.05
$FeOOH(s) + HCO_3^- + 2H^+ + e^-$	$= FeCO_3(s) + 2H_2O$	−0.05
$SO_4^= + 9H^+ + 8e^-$	$= HS^- + 4H_2O$	−0.22
$CO_2 + 8H^+ + 8e^-$	$= CH_4 + 2H_2O$	−0.24

Table 4   Molecular Composition of a Bacterial Cell

Element	Dry weight (%)
Carbon	50
Oxygen	20
Nitrogen	14
Hydrogen	8
Phosphorus	3
Sulfur	1
Potassium	1
Sodium	1
Calcium	0.5
Magnesium	0.5
Chlorine	0.5
Iron	0.2
Others	~0.3

The other major nutrients required by the microorganisms are nitrogen and phosphorous (Table 4). The three main forms of nitrogen found in microorganisms are proteins, microbial cell wall components, and nucleic acids. The most common sources of inorganic nitrogen are ammonia and nitrate. Ammonia can be directly assimilated into amino acid. When nitrate is used, it is first reduced to ammonia and is then synthesized into organic nitrogen forms.

Phosphorus in the form of inorganic phosphates is used by microorganisms to synthesize phospholipids and nucleic acids. Phosphorous is also essential for the transfer of energy during organic compound degradation.

Micronutrients are also required for microbial growth. There are several micronutrients that are universally required such as sulfur, potassium, magnesium, calcium, and sodium. Sulfur is used to synthesize two amino acids, cysteine and methionine. Inorganic sulfate is also used to synthesize sulfur-containing vitamins (thiamin, biotin, and lipoic acid) (Brock, 1979). Several enzymes including those involved in protein synthesis are activated by potassium. Magnesium is required for activities of many enzymes, especially phosphate transfer and functions to stabilize ribosomes, cell membranes, and nucleic acids. Calcium acts to stabilize bacterial spores against heat and may also be involved in cell wall stability.

Additional micronutrients commonly required by microorganisms include iron, zinc, copper, cobalt, manganese, and molybdenum. These metals function in enzymes and coenzymes and, with the exception of iron, are also considered heavy metals that can be toxic to microorganisms.

All of these factors are necessary to maintain the metabolic processes of microoganisms. One component of any biological design is providing the nutrients required by the microorganism. The carbon source should be the limiting factor in the biochemical reaction.

Environmental Factors

There are several environmental factors that can also affect biochemical reactions. These factors can control the type of bacteria that are prominent in the degradation of organics and will affect the rate of degradation. The main environmental effects are temperature, water, and pH. We will also review the different factors that can lead to toxic or inhibiting conditions.

Temperature — An Important Microorganism Growth Factor

As the temperature rises, chemical and enzymatic reaction rates in the cell increase. For every organism there is a minimum temperature below which growth no longer occurs, an optimum temperature at which growth is most rapid, and a maximum temperature above which growth is not possible. The optimum temperature is always nearer the maximum temperature than the minimum. Temperature ranges for microorganisms are very wide. Some microorganisms have optimum temperatures as low as 5 to 10°C; others as high as 75 to 80°C. The temperature range in which growth occurs ranges from below freezing to boiling.

No single microorganism will grow over this entire range. Bacteria are frequently divided into three broad groups as follows: (1) thermophiles, which grow at temperatures above 55°C; (2) mesophiles, which grow in the midrange temperature of 20 to 45°C; and (3) psychrophiles, which grow well at 0°C. In general, the growth range is approximately 30 to 40 degrees for each group. Microorganisms that grow in terrestrial and aquatic environments grow in a range from 20 to 45°C. Figure 3 demonstrates the relative rates of reactions at various temperatures. As can be seen, microorganisms can grow in a wide range of temperatures.

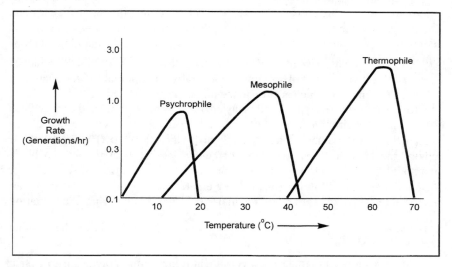

Figure 3   Temperature effect on growth rate. (From Brock, T.D., *Biology of Microorganisms*, Englewood Cliffs, NJ: Prentice-Hall, 1979. With permission.)

In general, biological reactions will occur year-round in the aquifer. Rates will be faster in the southern part of the country due to higher temperatures. Biological reactions in surface soils also will be affected by temperature; biological surface remediation will probably not occur during the winter months in northern climates.

Water

Water is an important factor for biochemical reactions. In saturated and unsaturated conditions, the bacteria may have to expend energy in order to acquire the water that they require. In the aquifer, the availability of water to microorganisms can be expressed in terms of water activity, which is related to vapor pressure of water in the air over a solution (relative humidity). Water activity in freshwater and marine environments is relatively high and lowers with increasing concentrations of dissolved solute (Brock, 1979). Bacteria can grow well in the salt water of an ocean (or 3.5% dissolved solids). Therefore, ground water, even from brine aquifers, will not pose any problems for bacterial growth.

In soil, water potential is used instead of water activity and is defined as the difference in free energy between the system under study and a pool of pure water at the same temperature and includes matrix and osmotic effects. The unit of measurement used is the *MPa*. As with water activity, this determines the amount of work that the cell must expend to obtain water. Generally, activity in soil is optimal at –0.01 MPa (or 30 to 90% of saturation) and decreases as the soil becomes either waterlogged near zero or desiccated at large, negative water potentials (Paul and Clark, 1989).

pH

Microorganisms have ideal pH ranges that allow growth. Within these ranges, there is usually a defined pH optimum. Generally, the optimal pH for bacteria is between 6.5 and 7.5; which is close to the intracellular pH. A bacteria cell contains approximately 1000 enzymes, and many are pH dependent (Paul and Clark, 1989). Most natural environments have pH values between 5 and 9. Only a few species can grow at pH values of less than 2 or greater than 10 (Brock, 1979). In environments with pH values above or below optimal, bacteria are capable of maintaining an internal neutral pH by preventing H^+ ions from leaving the cell or by actively expelling H^+ ions as they enter. Once again, the most important factor in regard to pH is to not allow major switches in pH during a remediation.

Toxic Environments

Many factors can render an environment toxic to microorganisms. Physical agents such as high and low temperatures, sound, and radiation and chemical agents such as heavy metals, halogens, and oxidants can inhibit microbial growth.

Chemical agents such as heavy metals and halogens can disrupt cellular activity by interfering with protein function. Mercury ions combine with SH groups in proteins, silver ions will precipitate protein molecules, and iodine will iodinate proteins containing tyrosine residues preventing normal cellular function. The effects of various metals in soil have been described (Dragun, 1988b) and are affected by the concentration and pH of the soil. Oxidizing agents such as chlorine, ozone, and hydrogen peroxide oxidize cellular components, destroying cellular integrity. Some *in situ* methods use hydrogen peroxide as an oxygen source; these methods may cause destruction of cells at higher concentrations of hydrogen peroxide.

MICROBIAL BIODEGRADATION OF XENOBIOTIC ORGANIC COMPOUNDS

The susceptibility of a xenobiotic compound to microbial degradation is determined by the ability of the microbial population to catalyze the reactions necessary to degrade the organics. Readily degradable compounds have existed on earth for millions of years; therefore, there are organisms that can mineralize these compounds. Industrial chemicals have been present on earth for a very short time on the evolutionary time scale. Many of these compounds are degradable, and many are persistent in the environment. Some xenobiotic compounds are very similar to natural compounds and bacteria will degrade them easily. Other xenobiotic compounds will require special biochemical pathways in order to undergo biochemical degradation.

A few definitions would be helpful here in order to understand different levels of biological reactions. Biodegradation means the biological transformation of an organic chemical to another form with no extent implied (Grady, 1985). Biodegradation does not have to lead to complete mineralization, which is the complete oxidation of an organic compound to carbon dioxide. Recalcitrance is defined as the inherent resistance of a chemical to any degree of biodegradation, and persistence means that a chemical fails to undergo biodegradation under a defined set of conditions (Bull, 1980). This means that a chemical can be degradable but, due to environmental conditions, the compound(s) may persist in the environment. With proper manipulation of the environmental conditions, biodegradation of these compounds can be demonstrated in laboratory treatability studies and transferred to field implementation.

Many articles have been written and many companies have promoted the use of specialized bacteria for the degradation of xenobiotic compounds. It is the opinion of the authors of this book that specialized bacteria will not have a wide application for *in situ* remediations. The reasons go beyond the standard argument that single microorganisms grown under laboratory conditions will not be successful under field conditions. *In situ* remediations will also require the delivery of the microorganism to the contaminant. As discussed in Chapter 1, the delivery will be dependent upon one of the carriers and the diffusion limitations associated with that carrier.

There are many natural processes that the microorganisms employ to expand the type of compounds that they can use as a food and energy source. We can create environments and provide growth factors that facilitate these processes. The rest of this section will discuss these various processes.

Gratuitous Biodegradation

Enzymes are typically described as proteins capable of catalyzing highly specific biochemical reactions. Enzymes are more specific to organic compound functional groups than to specific compounds. As Grady (1985) described, an enzyme will not differentiate between a C–C bond in a benzene molecule vs. a C–C bond in a phenol molecule. The functional capability of enzymes depends on the specificity exhibited towards the organic compound. A major enzymatic mechanism used by bacteria to degrade xenobiotic compounds has been termed "gratuitous biodegradation" and includes existing enzymes capable of catalyzing a reaction towards a chemical substrate.

In order for gratuitous biodegradation to occur, the bacterial populations must be capable of inducing the requisite enzymes specific for the xenobiotic compound. Often times this occurs in response to similarities (structural or functional groups) with naturally occurring organic chemicals. For example, a bacteria is producing the enzymes for benzene degradation. Chlorobenzene is introduced and is not recognized by the bacteria (its presence will not induce an enzyme to be produced); however, the enzymes already produced for benzene will also catalyze the degradation of chlorobenzene.

The capability of bacterial populations to induce these enzymes depends on structural similarities and the extent of substitutions on the parent compound. Generally, as the number of substitutions increases, biodegradability decreases unless a natural inducer is present to permit synthesis of required enzymes. To overcome potential enzymatic limitations, bacteria populations often induce a series of enzymes that modify xenobiotic compounds in a coordinated manner. Each enzyme will modify the existing compound such that a different enzyme may be specific for the new compound and capable of degrading it further. Eventually, the original xenobiotic compound will not be present. The compound will resemble a natural organic compound and enter into normal metabolic pathways. This concept of functional pathways is more likely to be completed through the combined efforts of mixed communities than by any one single species.

Cometabolism

Cometabolism has recently been defined as "the transformation of a non-growth substrate in the obligate presence of a growth substrate" (Grady, 1985). A non-growth substrate cannot serve as a sole carbon source to provide energy to support metabolic processes. A second compound is required to support biological processes allowing transformation of the non-growth substrate. This requirement is added to make a distinction between cometabolism and gratuitous biodegradation.

Microbial Communities

Complete mineralization of a xenobiotic compound may require more than one microorganism. No single bacteria within the mixed culture contains the complete genome (genetic makeup) of a mixed community. The microorganisms work together to complete the pathway from organic compound to carbon dioxide. These associations have been called consortia, syntrophic association, and synergistic associations and communities (Grady, 1985). We need to understand the importance of the microbial community when we deal with actual remediations. Conversely, we need to understand the limitations of laboratory work with single organisms; this work does not represent the real world of degradation. Reviewing the strengths of the communities will also reveal the limitations of adding specialized bacteria that have been grown in the laboratory.

Community Interaction

Microbial communities are in a continuous state of flux and constantly adapting to their environment. Population dynamics, environmental conditions, and growth substrates continually change and impact complex interactions between microbial populations. Even though environmental disturbances can be modified by microorganisms, microbial ecosystems lack long-term stability and are continually adapting (Grady, 1985). It is important to understand the complexities and interactions within an ecosystem to prevent failure when designing a biological system.

Communities and Adaptation

Mixed communities have greater capacity to biodegrade xenobiotic compounds due to the greater genetic diversity of the population. Complete mineralization of xenobiotic compounds may rely on enzyme systems produced by multiple species. Community resistance to toxic stresses may also be greater due to the likelihood that an organism can detoxify the ecosystem.

Community adaption is dependent upon evolution of novel metabolic pathways. As described by Grady (1985), a bacterial cell considered in isolation has a relatively limited adaptive potential, and adaption of a pure culture must come from mutations. Mutations are rare events. These mutations are generally responsible for enzymes that catalyze only slight modifications to the xenobiotic compound. An entire pathway can be formed through the cooperative effort of various populations. This is due to the greater probability that an enzyme system exists capable of gratuitous biodegradation within a larger gene pool. This genetic capability can then be transferred to organisms lacking the metabolic function which enhances the genetic diversity of the population. Through gene transfer, individual bacteria have access to a larger genetic pool allowing evolution of novel degradative pathways.

Genetic Transfer

Genes are transferred throughout bacterial communities by three mechanisms called "conjugation", "transformation", and "transduction" (Brock, 1979; Stanier et al., 1976; Moat, 1979; Grady, 1985; Rittman et al., 1990). Conjugation, apparently the most important mechanism of gene transfer in the natural environment, involves the transfer of DNA from one bacteria to another while the bacteria are temporarily joined. The DNA strands that are transferred are separate from the bacterial chromosomal DNA and are called plasmids (Brock, 1979; Stanier et al., 1976; Moat, 1979; Rittman et al., 1990). Plasmids exist in cells as circular, double-stranded DNA and are replicated during transfer from donor to recipient. Unlike chromosomal DNA which encodes for life-sustaining processes, plasmid genes encode for processes that enhance growth or survival in a particular environment. Examples of functions that are encoded on plasmids include antibiotic resistance, heavy metal resistance, and certain xenobiotic degradation enzymes (i.e., toluene) (Rittman et al., 1990).

Halogenated Hydrocarbons

There are many pathways available for the degradation of halogenated hydrocarbons. Some of the compounds degrade under anaerobic conditions; some require a co-metabolite for degradation. Many investigators have reported the mechanism of halogenated hydrocarbon transformation; Figure 4 is a compilation of several investigators' research into degradation of common halogenated hydrocarbon pollutants (Vogel and McCarty, 1987a; McCarty, 1986; Vogel and McCarty, 1987b; Kleopfer et al., 1985; Parsons et al., 1984; Barrio-Lage et al., 1986; Cooper et al., 1987; Wood et al., 1985).

DEGRADATION RATE

Microbial degradation of organic material is generally described as the time necessary to transform the substrate from its original form to another form. The final form can be a structurally different compound or complete mineralization into carbon dioxide, water, oxygen, and other inorganic matter. Biodegradation rates can be measured by the loss of the original substrate or the consumption of oxygen or other electron acceptors.

There are three factors that control the rate of degradation in an *in situ* remediation (assuming that the bacteria are present and the contaminants are degradable): (1) the rate of organic entering the area of the aquifer, (2) the rate that the final electron acceptor enters the area, or (3) the rate that the bacteria can use the final electron acceptor.

The original location in which the organic comes into contact with the aquifer is usually the area of highest concentration. If non-aqueous phase liquids (NAPLs) are present, this area will be the source for the plume that will be created. In these areas of high concentration, the final electron acceptor will control the rate of degradation.

Oxygen and nitrate can be considered to be used instantaneously (in comparison to the rate of movement by advection) in areas of high organic concentration. These compounds will be consumed at the rate at which they enter the highly contaminated area (Wiedemeier et al., 1994). Microorganisms will use sulfate, iron III (ferric iron), and carbon dioxide at a rate that is limited by the biochemical reaction rate.

The organic will move by advection while the plume is spreading. When the mass movement of the contaminant by advection is equal to the supply rate of the final electron acceptors, the plume will cease to move further. Where the plume stops, the rate of organic entering the area will be the rate of degradation. This equilibrium can also occur when the plume is being remediated. The aquifer will test clean when the rate of diffusion from the relatively stagnant zones of

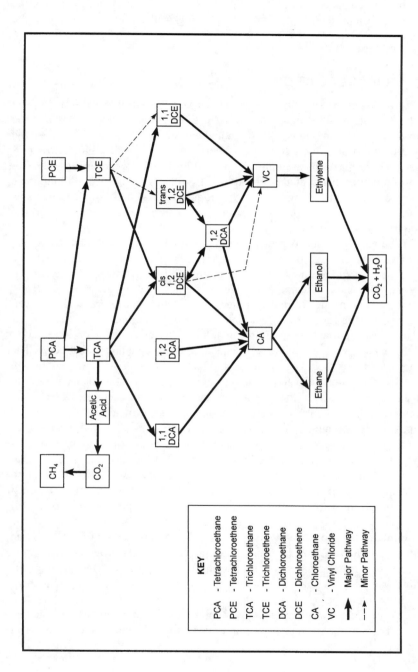

Figure 4 Transformations of chlorinated aliphatic hydrocarbons.

the aquifer (see Chapter 1 for a full discussion) is equal to the rate of degradation by the bacteria. This equilibrium may prevent the compounds from reaching the monitoring system, depending on the monitoring well placement relative to the plume.

IN SITU BIOREMEDIATION

The next step to applying biological reactions to the remediation of a site is understanding how the bacteria interact with the natural environment and the manmade organic compounds. As we stated earlier, the microorganisms use the organic material as a food source for energy and as building blocks for new bacteria. These reactions are natural, and full bacterial populations are distributed throughout unsaturated and saturated soil. The presence of manmade organics can shift the dynamics of the population, with the microorganisms best able to use the new organics as the food source outcompeting the other microorganisms.

This means that the organic compounds that have been released to the environment are undergoing natural or "intrinsic" bioremediation. For years we thought that in situ bioremediation was controlled by the transfer of oxygen (or nitrates) to the area of contamination. We realize now that oxygen represents only the fastest reaction rate. In reality, other final electron acceptors, while slower, can be responsible for more of the contaminant degradation.

The most important step to implementing in situ bioremediation is to understand the natural reactions and the subsequent rates of remediation that occur. The best term for these processes is "intrinsic bioattenuation". The next section will discuss these reactions and the sampling and analysis methods that are used to delineate these processes. Once we understand the natural reactions, we can then determine if it is necessary to increase the natural rates of reaction.

Intrinsic bioattenuation processes can actively control the migration of and, in many cases, remediate many organically contaminated soils and groundwater plumes. This control and/or remediation is the result of combining the several different biochemical reactions that we have already discussed. In order to understand which reactions are contributing to the overall remediation, we must gather data on the various microbial processes. The biogeochemical data will give us the basis for delineating the intrinsic reactions that are occurring at the site. Summarizing the various reactions based upon this data will then provide the total assimilative capacity when combined with the hydrogeological data.

We have already discussed the role of hydrogeology and its thorough understanding as a basic component of any groundwater remediation strategy. Knowledge of the hydrogeology is very important in order to apply intrinsic bioattenuation to groundwater situations. The contaminant plume must be adequately delineated. Once the groundwater plume has been delineated, biogeochemical data can be collected.

Biogeochemical Characterization

In most contaminated groundwater situations, considerable sampling and analytical data concerning contaminant levels and contaminant types already exists. However, biogeochemical data is seldom available. We need this data in order to understand what is occurring below ground. By collecting data on the chemicals that are involved in the biochemical reactions, we can begin to understand the intrinsic reactions.

One limitation is that there are no absolute values that can be used when analyzing the data at a site. The designer must compare the upgradient, impacted, and downgradient values of each of these groups of compounds. Natural bacterial activity and geologic conditions change so dramatically between sites that only the on-site comparison of these biogeochemical parameters can be used to analyze intrinsic reactions.

An example of this is the changing proportions of BTEX constituents at a fuel-contaminated site and basic knowledge of the individual components and their susceptibility to aerobic and anaerobic biodegradation. Benzene tends to persist under some anaerobic conditions and thus tends to become the predominant component under some anaerobic, but not necessarily all, conditions. Conversely, benzene tends to biodegrade relatively faster than the other constituents under an aerobic scenario. The change in relative concentrations between the benzene, toluene, ethyl benzene, and xylenes across the site will indicate what type of reactions are occurring. Chlorinated organics can be subject to anaerobic biodegradation and certain degradation by-products or lack of them can provide important information as to the possibility of intrinsic bioattenuation of the contaminants. After collecting preliminary information on organic destruction and by-products, the designer needs further data to confirm and completely understand the specific reactions that are occurring.

Biogeochemical data that we will need to characterize the contaminated area can be broken into two main groups based upon the sampling method and the way the information will be utilized. Down-the-well parameters will provide general environmental conditions, and chemical analyses of the ground water will provide the specific data needed to delineate the specific biochemical reactions.

Down-the-Well Parameters

Environmental conditions must be understood at this site. First, the conditions provide important indicators of the reactions occurring. Second, many of the reactions will only occur under certain environmental conditions. For example, certain anaerobic reactions will only occur under specific redox conditions. There are a number of groundwater instrument vendors that provide various multiparameter probes that can be physically lowered into 2-in. and larger diameter monitoring wells. The primary parameters of interest, are:

- Dissolved oxygen
- Redox potential
- pH
- Temperature
- Specific conductance

It is important to collect the down-the-well data at different depths over the entire screened interval. Aerobic or anaerobic conditions may change with depth. Also, there may be zones of aerobic/anaerobic interaction. These probes and their associated membranes should not be subjected to free product, even with methods that try to protect the probe. Consideration should also be given to gathering the down-the-well parameters both before and after purging of the monitoring wells. Rapid evacuation of ground water from a monitoring well can result in cascading of ground water down the screen. Exposure to the atmosphere can cause erroneous results for dissolved oxygen and other parameters. Some recent discussions have also suggested leaving the multiparameter probes in the well during purging to better characterize the ground water conditions. Pump and flow through cells can replace down-the-well probes. While the accuracy may be greater, they will require more time and effort to produce the data.

Chemical Analysis

The chemical analysis provides specific information on the electron acceptors being utilized. These data, when combined with the hydrogeologic data, provide rates of reaction and are the basis for modeling the activity at the site.

Analysis of primary electron acceptors is used mainly to determine the presence and utilization of these compounds. The primary electron acceptor for aerobic metabolism is oxygen, the presence of which can be estimated by down-the-well probe methods and verified via dissolved gas determination. As many impacted aquifers are anoxic/anaerobic due to the depletion of oxygen from the impact of the contaminants and other geochemical factors, the electron acceptors for anaerobic metabolism must be investigated and understood. As discussed before, the main anaerobic electron acceptors are nitrate, iron III, sulfate, and carbon dioxide (in order of increasing reducing conditions).

In addition to evaluating the utilization of electron acceptors, the presence of dissolved biogenic gases should be determined. The purpose of these data is to confirm biochemical activity. For example, during aerobic respiration, bacteria use O_2 as a final electron acceptor and produce CO_2. However, most contaminated soils and aquifers do not have O_2 available. This is a result of bacterial activity consuming available O_2 as organics are degraded to produce energy and new bacteria. Once available O_2 is consumed, facultative aerobic and anaerobic bacteria use other final electron acceptors such as nitrate, sulfate, CO_2, or organics to complete respiration processes. Just like O_2 and CO_2, these compounds can be used to detect and understand biochemical reactions.

Nitrate, nitrite, ammonium, and molecular nitrogen are part of the total nitrogen cycle involved in intrinsic bioattenuation. The oxidized forms of nitrogen are utilized, and reduced forms are produced during degradation of the contaminants. In addition, nitrate can be reduced to ammonia, which can serve as a nitrogen source (nutrient) for microbial metabolism. It can also be further reduced to molecular N_2, which can be "fixed" by some indigenous microbes and utilized as a nitrogen source or utilized by vegetation if the ground water is in close association with the root zone (rhizosphere).

Elevated concentrations of iron III are often present in impacted groundwater plumes. These concentrations are often the result of the biochemical reduction of iron II (ferrous iron) to ferric iron, described earlier. Therefore, both ferrous and ferric iron data should be collected to determine if the presence of iron in ground water is indigenous or perhaps biologically produced. In addition to evaluating if ferrous iron is biologically produced, these data are also useful for the design, operation, and maintenance of remedial equipment because high concentrations of iron can foul remedial equipment such as air strippers and recovery wells, or foul geological formations.

Sulfate (SO_4^{-2}) reduction by microbes in impacted groundwater can also occur. Highly active sites will form H_2S, the source of the "rotten egg" smell. However, H_2S will react with metals (Fe II, Mn II, and others) to form metal sulfides. Therefore, H_2S in ground water may not be present at significant levels. Benzene, toluene, ethyl benzene, xylenes, naphthalenes, etc., and many chlorinated compounds have been shown to be biologically degradable under sulfate-reducing conditions (EPA, 1993). Sulfate reduction is a very important biogeochemical process in contaminated ground water. Due to the interaction of the reduced form of sulfide with other compounds, only data on the level of SO_4^{-2} can be used to estimate the rate of bio-activity.

Carbon dioxide, a by-product of aerobic metabolism and fermentative processes can be reduced to methane by *in situ* intrinsic bioattenuation processes (methanogenic) and can be analyzed as a dissolved gas in ground water. Methanogenesis is an important biogeochemical process as BTEX constituents have been shown to be susceptible to biodegradation under methanogenic conditions (Howard et al., 1991). Many chlorinated organics can also be biologically degraded via methanogenic processes. Methane, a by-product of carbon dioxide reduction, can be utilized by methanotrophic microbes as a carbon source in the presence of dissolved oxygen; chlorinated organics and other organic contaminants can be cometabolized under methanotrophic conditions (Nyer, 1992).

Organic compounds also can be useful in tracking intrinsic bioreactions. For instance, investigation of intrinsic bioattenuation of chlorinated organics under reducing conditions may be better understood by adding analyses for various organic acid byproducts (acetate, formate) and various end products, such as ethane, to help verify various anaerobic pathways.

One other group of groundwater parameters can be useful in assessing intrinsic bioattenuation and arriving at a remedial strategy. Biological oxygen demand, chemical oxygen demand, and total organic carbon, can be used to monitor the general organic content in the aquifer. Since the bacteria will utilize all of the organics available, it is useful to know how much organic is present that was not detected in the specific organic analysis.

A list of parameters that should be considered when evaluating intrinsic bioremediation is shown in Table 5. Several good reference articles for analytical methods suitable for biogeochemical characterization are available (Wiedemeier et al., 1994; Norris et al., 1993).

Table 5   Evaluation of Natural/Intrinsic Bioattenuation

Down-the-well parameters (after purging):
Dissolved O_2
Redox potential
pH
Temp (˚C)
Specific conductance

Laboratory parameter	**Analytical method**
Target compound list VOCs	CLP SOW, OLM 1.9[a]
Nitrate	Method 353.2
Nitrite	Method 353.2
Nitrogen, ammonia	Method 350.3
Manganese (total)	Method 6010A
Manganese (dissolved)	Method 6010A
Iron (total)	Method 6010A
Iron (dissolved)	Method 6010A
Iron (ferrous)	Method 315B
Sulfate	Method 375.4
Sulfide	Method 376.1
Total organic carbon	Method 415.1
Chlorides	Method 325.2
Light hydrocarbon scan (Ethane, ethene)[b]	
Permanent Gases (carbon dioxide, oxygen, nitrogen, methane and carbon monoxide)[b]	

Optional groundwater parameters
Chemical Oxygen Demand (COD)
Biochemical Oxygen Demand (BOD)
Total Organic Carbon (TOC)

[a] U.S. Environmental Protection Agency Contract Laboratory Program, Statement of Work for Organic Analysis OLM 1.9, July 1993.
[b] Refers to U.S. Environmental Protection Agency SW 846.

Interpretation of Biogeochemical Data

Once the data are gathered, interpretation of the *in situ* biogeochemical processes can commence. It is recommended that the focus of this interpretation be based on observation of biogeochemical process trends due to the very nature of *in situ* geology and the complex interactions of the biogeochemistry.

Oxygen consumption, denitrification, iron reduction, sulfanogenesis, and methanogenesis are the principle biological processes responsible for intrinsic bioattenuation of organically contaminated aquifers. The biogeochemical data can be used to show specific reactions that have occurred at the site. The changes in concentrations across the site can be combined with the hydrogeologic information, and mass removals can be calculated.

Wiedemeier et al. (1994) summarized these functions for a site at Hill Air Force Base in Utah. Stoichiometric relationships for oxygen, denitrification, iron reduction, sulfanogenesis, and methanogenesis in relation to BTEX compounds were analyzed and the assimilative capacity of the site was developed.

The investigation team analyzed each of the electron acceptors that would be available at the site. Then they assumed a stoichiometric reaction for each electron acceptor based upon the complete degradation for BTEX compounds. Table 2 provides the stoichiometric reactions for each electron acceptor. Based upon similar reactions, Weidemeir et al. calculated that each electron acceptor would be consumed while degrading the following amounts of BTEX;

3.14 µg	O_2 to metabolize	1 µg	BTEX
4.9 µg	NO_3^- to metabolize	1 µg	BTEX
21.8 µg	Fe^{+3} to metabolize	1 µg	BTEX
4.7 µg	SO_4^{-2} to metabolize	1 µg	BTEX
0.78 µg	CH_4 to metabolize	1 µg	BTEX

The site had the following background concentrations of each electron acceptor:

6.0 µg/L	O_2
17.0 µg/L	NO_3^-
50.5 µg/L	Fe^{+3}
97.6 µg/L	SO_4^{-2}
2.04 µg/L	CH_4

The total assimilative capacity of the site groundwater was calculated to be

Electron acceptor	BTEX assimilative capacity (µg/L)
O_2	1,920
NO_3	3,570
Fe^{+3}	2,300
SO_4^{-2}	20,500
CH_4	2,600
Total	30,890

The highest observed total BTEX concentration found at the site was 21,475 µg/L. Since this was close to the theoretical limit of BTEX in ground water from a JP-4 source, it was concluded that the assimilative capacity of the ground water exceeds the expected influx of BTEX (21,475 µg/L) and should prevent the plume from migrating.

Many assumptions went into the above calculations; however, the conclusions can be confirmed with the field data. Analysis of the monitoring data over time shows that the plume is not moving, and the biogeochemical data shows that the electron acceptors are being utilized. Combining all of this information, one can make a strong conclusion that biological activity is responsible for the destruction of the BTEX compounds.

Stoichiometric relationships for other organic contaminants are also available in the literature, and of particular interest are stoicmetric relationships and kinetics derived from *in situ* field experiments. Recent findings of the assimilative capacities of anaerobic bioattenuation processes, which are often active within an impacted plume of ground water, reveal that these anaerobic processes are often the most active at mass removal of the contaminants. Norris (1993) reviewed the reactions for chlorinated organic compounds and it is suggested that the reader refer to that source directly if more detailed information is required in this area.

The reader can now see why all of the biogeochemical data are necessary. These data must be collected in order to compare biologically mediated changes across the impacted groundwater plume. Remedial monitoring of organic constituents will confirm that they are not migrating. Combining all of the data leads to the strong conclusion that intrinsic biological activity at a site is capable of treating contaminated groundwater. If data are required regarding the rate of degradation and its effect on the plume, then models can be used as a predictive tool.

Modeling Support for Intrinsic Bioattenuation

Biochemical degradation is not the only process that is affecting the concentrations of the organic compounds. Hydrogeological factors such as retardation, diffusion, and dilution will also have a dynamic affect on the concentration Therefore, in order to assess intrinsic bioattenuation one must include all of the factors that can result in decreasing contaminant levels. Several groundwater models have been developed for this purpose. When the groundwater model is used for tracking and predicting biochemical reactions, it can be referred to as a "biomodel". With the use of biomodeling, intrinsic bioattenuation can be better visualized and comprehended. The sophistication of the biomodeling should be dictated by consideration of the following:

1. Can modeling be used to reduce monitoring requirements?
2. Are there any recommended models that a regulator may desire to use, and are they appropriate for this particular situation?
3. What is the most appropriate model to fit the site requirements and biogeochemical processes?
4. What are the economics of the modeling exercise in relation to the site remedial objectives?

Most solute transport models in use today are a combination of groundwater flow models that describe a velocity field, recharge/discharge, and changes in aquifer storage for site-specific conditions and transport models, which may address dispersion, density-dependent flow, or some reversible or irreversible chemical phenomena. Most solute transport models do not address biochemical reactions. Reversible equilibrium and controlled sorption is typically simulated by use of a retardation factor or coefficient. The following section describes models that include biogeochemical reactions along with the other flow and transport factors. Table 6 summarizes the models and the contact address for more information.

Table 6   Models That Can Incorporate Biochemical Reactions

Model name	Release date	Contact
BIOPLUME II	1987	H.S. Rifai; Rice University, Dept. of Environmental Sciences and Engineering, P.O. Box 1892, Houston, TX 77251
SLAEM	1992	O.D.L. Strack; University of Minnesota, Dept. of Civil Engineering, Minneapolis, MN 55455
AT123D	1987	G.T. Yeh; Dept. of Civil Engineering, Pennsylvania State University, University Park, PA 16802, or International Ground Water Modeling Center, Colorado School of Mines, Golden, CO 80401
PORFLO	1985	N.W. Kline; Boeing Computer Services, P.O. Box 300, Richland, WA 99352
MT3D	1992	C. Zheng; S.S. Papadopulos & Assoc., Inc., 7944 Wisconsin Ave., Bethesda, MD 20814

Available Models That Can Incorporate a Biodegradation Factor

There are several models that can be set up to include biochemical reactions. Most of the models simply have a factor that accounts for disappearance of the compound. The model does not distinguish between adsorption, destruction, or any other process that would remove the organics from the carrier. This may limit the accuracy of the models, but they are still very useful in understanding and predicting what is going on below ground.

The most recognized model specifically developed for biochemical reactions is BIOPLUME II, which is a two-dimensional solute transport model used to compute changes in concentration over time due to advection, dispersion, mixing, and retardation. The model simulates the transport of dissolved hydrocarbons under influence of oxygen-limited biodegradation. It also simulates reaeration and anaerobic biodegradation as a first order decay in hydrocarbon concentrations. BIOPLUME II is based on the United States Geological Survey (USGS 2D) solute transport model MOC. It solves the transport equation twice: once for hydrocarbon and once for oxygen. The model assumes an instantaneous reaction between oxygen and hydrocarbon. It can simulate natural biodegradation processes, retarded plumes, and in situ bioremediation schemes. At the present time, BIOPLUME III, a three-dimensional version of BIOPLUME II, is being developed.

The other models listed in Table 6 incorporate some type of disappearance factor. The text may describe this factor as decay, retardation, or in several other terms. It is important to understand that these losses are dependent upon biochemical reactions. If a model has been previously selected, the designer should make sure that the disappearance due to biochemical reactions is included during setup and operation. If a model has not been selected, then the importance of the biochemical reactions should be included in the model selection. It is hoped that in the future more models will incorporate methods that can account for the various biochemical reactions.

Risk/Biomodeling Approach

The last, but certainly not least, important part of an intrinsic bioattenuation remedial strategy is the incorporation of risk factors. Declining risk values within reasonable time frames is one of the most important aspects of establishing the viability of intrinsic bioattenuation at a contaminated site. It is important to shown that the site provides no risk to human health or environment while it is undergoing intrinsic remediation. Since intrinsic bioremediation is considered a relatively slow process, risk assessment becomes an important tool.

One of the early problems with applying intrinsic bioremediation is that regulators considered the process as a "do nothing" alternative. It was felt that the owner of the property was simply trying to save money. As we have learned more about the various reactions that naturally occur, the destructive nature of the process has gained appeal. However, the process can be slow. The regulators may worry that people and the environment will be exposed for too long while the intrinsic processes are working.

The best way to understand and deal with this concern is to perform a risk assessment. The basic premise of the risk assessment is that the organic compound is not dangerous, but when the organic compound is in contact with a human there is danger. A risk assessment studies all of the pathways by which a human may come into contact with the organic compound. Different risk levels are assigned based upon the exposure route. If the exposure route can cause harm, then that pathway should be controlled. For example, a gasoline spill can create vapors that enter the basement of a house. While natural reactions can eliminate the compounds over time, the immediate risk for inhalation of the vapors is not acceptable. In this case, a vapor extraction system (see Chapter 4) may be used to reduce the vapors and source of vapors to the point where the risk of exposure has been eliminated. At that point, intrinsic remediation can treat residual impacts.

Not all risks are immediate. Risks of cancer usually occur over extended periods of time. In these cases, the bioremediation must be included in the risk assessment. The bioremediation will result in a continuous decrease in the concentration of the organic. The risk assessment must, therefore, incorporate this change in concentration over the time of exposure. The assessment should not be based on the initial concentration as the exposure level.

As can be seen, the risk assessment can become an integral part of the remediation evaluation and design. The same analysis used for human health can be repeated for ecological health.

Case Histories of Intrinsic Bioattenuation

Indiana Superfund Site

The Seymour site, located in Seymour, Indiana, is 13 acres in area. The facility operated from 1971 to 1980, at which time the U.S. Environmental Protection Agency (EPA) shut down the site because of releases to the air and surface water. After approximately 60,000 55-gallon drums and 100 tanks were removed by the EPA as part of its emergency response, the agency employed a subcontractor to conduct a remedial investigation and feasibility study (RI/FS), which focused on defining the extent of soil and groundwater contaminant concentration at the site.

A later evaluation of the actual concentrations found in the ground water and expected transport based on retardation indicate that the original model was overly conservative in its estimation of the transport of some compounds, particularly compounds subject to biodegradation. Table 7 is a list of some of the compounds in the groundwater plume and compares the distance each theoretically should have traveled. The distance is based upon entrance into the groundwater system in 1980 and that retardation is the only process that affects migration in the subsurface. The table compares the theoretical travel distances to those that have actually occurred as determined by analytical data from monitoring wells at the site.

Table 7   Comparison of Model and Actual Contaminant Migration

	Retardation factor	Model predicted travel distance beyond site boundary (ft)	Actual travel distance beyond site boundary (ft)
Tetrahydrofuran	1.0	4300	1450
1,4-Dioxane	1.0	4300	3900
Benzune	3.5	1250	100
Phenol	1.7	2500	0

Except for 1,4-dioxane, which has traveled almost as far as expected, the other compounds in Table 7 have migrated smaller distances than anticipated. Figures 5, 6, and 7 depict actual plume travel distances vs. theoretical travel distances based on transport modeling for benzene, phenol, and 1,4-dioxane, respectively. As can be seen in Table 7, tetrahydrofuran and 1,4-dioxane, which have identical retardation factors and similar solubilities in water, exhibit greatly different migration rates. 1,4-Dioxane traveled about 2.5 times farther than tetrahydrofuran. This is probably the result of the differing degradation rates of the two compounds. Tetrahydrofuran is amenable to biodegradation (although at a low rate), whereas 1,4-dioxane is not. In addition, benzene and phenol, which also are amenable to biodegradation should have traveled 1250 and 2500 ft, respectively, beyond the site boundary. The analytical data indicate that these two compounds have migrated less than 100 ft from the site boundary. These compounds are known to be degraded under aerobic and anaerobic conditions.

A biogeochemical characterization of this site verified that aerobic and anaerobic bioattenuation processes were present and actively degrading many of the contaminants. Utilization of this information was integral to formation of an altered remedial approach which included a reduced pump-and-treat scenario. It was decided that natural destruction was preferable to bringing the compounds above ground.

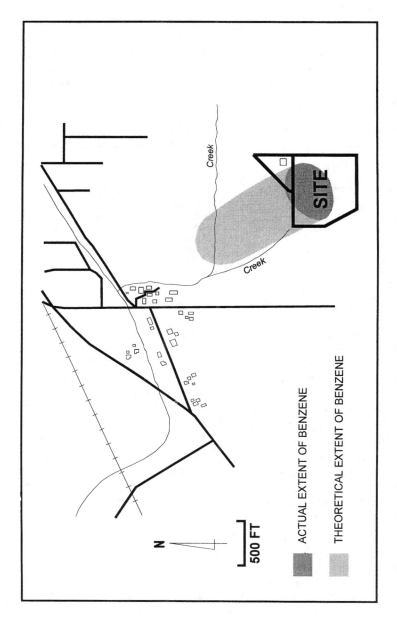

Figure 5   Theoretical and actual distributions of benzene in the shallow aquifer, Seymour site, Seymour, Indiana.

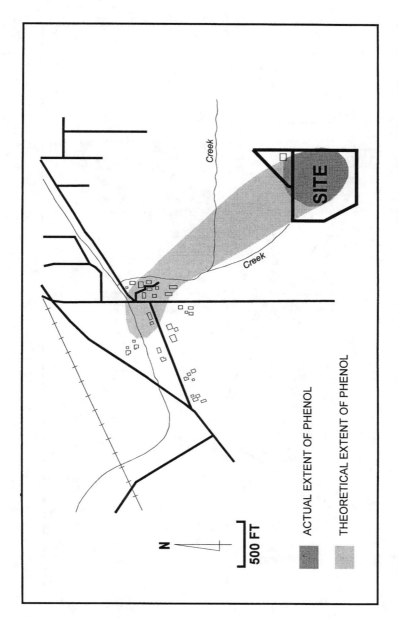

Figure 6 Theoretical and actual distributions of phenol in the shallow aquifer, Seymour Site, Seymour, Indiana.

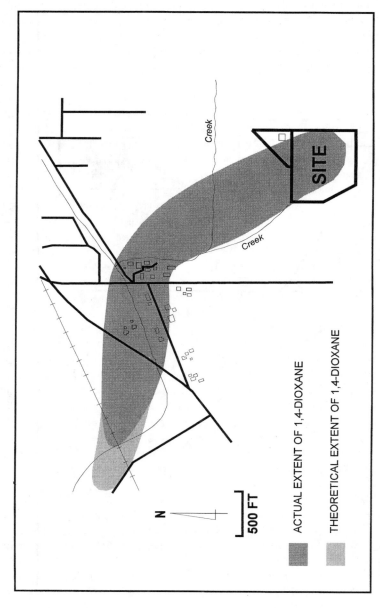

Figure 7   Theoretical and actual distributions of 1,4-dioxane in the shallow aquifer, Seymour site, Seymour, Indiana.

Michigan Landfill

During the 1960s and 1970s, the West KL Avenue Landfill in Kalamazoo, Michigan, was the repository for an estimated 5 million yd^3 of refuse and undetermined amounts of bulk liquid and drummed chemical waste. In 1979, the 87-acre site was closed permanently due to the discovery of contaminants in nearby residential drinking water wells. In 1983, the site was placed on the National Priority List due to the discovery of acetone, methyl ethyl ketone, methyl isobutyl ketone, dichloroethane, benzene, and other contaminants in ground water near the site.

The leachate plume was approximately 1/2 mile wide and 1/4 mile long and contained a number of priority pollutants, including benzene. The remedial investigation (RI) found that over the 8 years of monitoring the extent of the landfill plume, the plume had migrated no more than 200 ft. Flow and transport parameters determined in the RI indicated that the plume should have migrated several thousand feet during that period. This discrepancy was attributed to biodegradation processes within the plume by the RI report.

The Record of Decision (ROD) for this Superfund site included a Resource Conservation and Recovery Act (RCRA)-type landfill cap and a conventional pump-and-treat (fixed film biological treatment) approach. Because of regulatory restrictions on where the treated waters could be discharged, the resulting system was going to involve a mile-long pipeline buried below the frost depth to the nearest Publicly Owned Treatment Works (POTW). This remediation system was estimated to cost at least $33 million with an estimated operating period of 30 years.

Due to the magnitude of remedial costs and treatment duration specified in the ROD, it was decided to assess the extent to which the ongoing biodegradation processes could decrease these costs and minimize or eliminate the need for groundwater pumping. The first step was to develop and calibrate a biodegradation model utilizing the RI data supplemented with literature values. The model demonstrated that if the biodegradation processes are ignored and only the flow-and-transport processes are taken into account for that 30-year period, the plume would have extended up to $1^1/_2$ miles farther to the west-northwest, bringing the plume into contact with several residential wells in that area. Consequently, the biodegradation processes have had a major influence in containing and controlling the extent of the plume.

Next, three bioremediation alternatives were evaluated: the biodegradation implications of capping the landfill, bioventing by air extraction, and bioventing by air injection. The model predicted that, with the landfill cap in place, the ongoing natural biodegradation processes would decrease off-site contaminant concentrations to acceptable levels in approximately 20 years. The bioventing approaches were predicted by the biomodel to decrease remediation duration to less than 5 years and cost an estimated $11.8 million ($0.8 million for the bioventing and $11 million for the landfill cap). These are significant reductions in the estimated system that had already been agreed upon in the ROD for the site. The parties to the ROD have subsequently agreed to amend the remediation approach to include bioremediation. This site is now a EPA Bioinitiative Research Site and further research is being conducted into the total role of intrinsic bioattenuation processes at this site.

Hydrocarbon Sites

A major oil company was interested in reducing the financial liabilities associated with leaking underground storage tanks located throughout the U.S. and abroad by utilizing intrinsic bioattenuation in lieu of traditional active pump-and-treat systems. Recognizing the regulatory hurdles involved, they focused their initial efforts on the state of Florida, where they presently have several hundred sites involved in the state's Underground Storage Tank (UST) program. The oil company undertook a review of a number of sites in southeast Florida and selected four sites for evaluation which represented a cross section of the types and distribution of contamination typical of service station sites. At each of the four sites, assessment reports defining aquifer characteristics and delineation of the horizontal and vertical extent of hydrocarbon impacts had been approved by state regulatory agencies. According to state guidelines, the impacts at each site would require implementation of remedial actions. In fact, at one site remediation was underway, and the study was initiated to see if active remediation could cease.

The objectives of the study were to collect field and laboratory data to demonstrate the presence of natural, intrinsic bioattenuation, and to develop a model to evaluate whether intrinsic bioattenuation is a viable alternative to traditional groundwater remediation systems. Soil vapor, saturated soil, and groundwater data were collected from nonimpacted upgradient points, locations within the plume, and near the downgradient edge of the plume. In addition, saturated soil samples were evaluated at a treatability laboratory for the presence of heterotrophic and petroleum-degrading microbes. The evaluation indicated that the subsurface environment was amenable to biochemical activity and that sufficient populations of petroleum degrading microbes existed.

Using hydrogeologic data from site assessment reports, field measurements, and analytical data, the transport and biodegradation of the dissolved hydrocarbon plume was simulated with the BIOPLUME II model. A reasonable match between actual monitoring well analytical data and the simulated plume was obtained by calibrating the model with the longitudinal and transverse dispersivities, the source influx, and the reaeration coefficient. The model was then used to predict the future extent of the plume under the no-action alternative.

For one of the sites that was studied, the changes in the total benzene, toluene, ethyl benzene and xylenes (BTEX) concentration are presented in Table 8. The decrease in the BTEX concentrations are the result of biological degradation in combination with abiotic processes (e.g., volatilization, dilution, adsorption, retardation). Field measurements of dissolved oxygen and the oxidation-reduction potential (Eh) indicated that aerobic vs. anaerobic conditions were present due to the Eh values above +50mV (indicative of aerobic conditions) and the transition from high dissolved oxygen (DO) levels upgradient to lower levels in the impacted zone.

Table 8 Comparison of BTEX Concentrations

Well ID	High concentration	Date of sampling	8/5/93 concentration	Historical decrease (%)
MW-1	4110	3/89	10	99
MW-2	8246	9/91	1030	88
MW-5	189	3/89	64	67
MW-6	2094	9/90	280	87
MW-8	1030	5/89	21	98
MW-9	2940	5/89	172	94
MW-10	5140	5/89	1020	80
MW-11	3090	5/89	737	77
MW-12	365	7/89	10	97
MW-16	710	10/91	16	98

Note: All concentrations in parts per billion (ppb).

The biomodeling showed that intrinsic bioattenuation is a viable remedial alternative for UST sites throughout Florida and that this alternative is the preferred remedy at three of the four sites studied, in lieu of more costly pump-and-treat systems. At the fourth site, where remediation was already underway, the plume was already at the site boundaries and additional biomodeling will be performed to determine when active remediation can cease.

ENHANCED BIOREMEDIATION

There are two ways in which we can enhance bioremediation at impacted sites. The first way is to enhance the reactions that are already occurring at the site. The main method of encouraging existing reactions is to supply oxygen (or nitrate) to the zones that have depleted their natural oxygen content. The second way to enhance bioremediation is to introduce material that will change the environment or encourage production of biological enzymes that produce reactions that we desire. The second type of enhancement is usually centered around changing the environment to anaerobic conditions to encourage dehalogenation, or the introduction of cometabolites in order to produce enzymes that degrade target compounds.

Existing Reactions

As discussed in the previous section, bioremediation rates can be limited by the availability of the electron acceptor (reactions using O_2 and NO_3^-) and by the bacterial reaction rates (reactions using SO_4^{-2}, Fe, or organics). This means that if we want to increase the rate of bioremediation, we will have to rely on bacteria that use O_2 and NO_3^- as their final electron acceptors. While these electron acceptors provide faster reactions, they usually are not available in the quantities necessary to complete the remediation. Therefore, O_2 or NO_3^- must be delivered to the bacteria in order for them to maintain a high rate of reaction. Originally, all *in situ* bioremediations were based upon the delivery of O_2. Now that we have a better understanding of the other intrinsic reactions, bioremediations rely on a variety of reactions, but O_2 is still the best way to enhance the rate of remediation.

There are five main methods of supplying bacteria with oxygen or nitrate in the saturated zone:

1. Injection of aerated water
2. Injection of hydrogen peroxide
3. Injection of nitrate
4. Air sparging
5. Venting

The oldest method of supplying oxygen to bacteria in the aquifer is to saturate the water with oxygen before injecting it into the aquifer. This can be accomplished by standard aeration methods or by aerating the water with pure oxygen. Oxygen has a limited solubility, 7 to 10 mg/L, when standard aeration is used. The temperature will affect the solubility in either situation. The use of pure oxygen will increase the maximum solubility by a factor of about three. Therefore, highly oxygenated water is recirculated through the aquifer as fast as possible.

Hydrogen peroxide also can be used as an oxygen source. Hydrogen peroxide decomposes to form water and oxygen. Hydrogen peroxide is very soluble, and high concentrations (100 to 1000 mg/L) can be added to the water being injected into the aquifer. However, hydrogen peroxide is also a strong oxidant and can be harmful to the bacteria if too high a concentration is used.

Nitrate has also been used as a final electron acceptor for the bacteria. Once again, nitrate is very soluble in water and high concentrations can be supplied with the injection water. There are two main problems with nitrate. First, all bacteria cannot use nitrate as the final electron acceptor and not all compounds will degrade under denitrifying conditions. Second, nitrate is considered a contaminant and the design may have to include the removal of residual nitrate at the end of the cleanup.

Sparging can be used to add oxygen directly to the aquifer. Air is pumped into the aquifer through wells. Air is 21% oxygen, which transfers to the water in the aquifer. Ground water, passing through the area of the wells, is then supplied with oxygen. Sparging is a cost-effective source of oxygen and will be discussed in detail in Chapter 6.

Finally, oxygen can be supplied by venting the unsaturated zone and allowing the natural transfer of oxygen from the unsaturated zone to the water of the aquifer. Because of the limited surface area of the water, oxygen transfer is a slow process. But, when venting is used as part of the project, it can be an important aspect of the oxygen supply. The best way to know if venting will assist with oxygen transfer is to analyze the oxygen content of the gases in the vadose zone. Atmospheric oxygen content is about 21%. If the vadose zone gases are significantly below that concentration, then fresh air brought into the unsaturated zone will increase the rate of oxygen supply. The authors have worked on projects where the oxygen content of the unsaturated zone was less than 2%.

Venting was successfully applied at those locations, and the rate of biodegradation in the aquifer increased significantly.

All of these methods have advantages and disadvantages. As will be discussed later in this section, the geology must also be included in the selection process between the various O_2 and NO_3^- delivery methods. The design engineer will have to pick the right method for each particular site. Table 9 shows a cost comparison from the literature of the various methods of oxygen supply. These numbers should only be used as background information, and cost analysis should be completed for each unique site.

Change of Environment

The intrinsic bioremediation section of this chapter covered the natural reactions that controlled and remediated degradable compounds in an aquifer. Intrinsic remediations normally are associated with petroleum hydrocarbon organic compounds. Natural reactions also occur with chlorinated compounds. We discussed in the beginning of this chapter the natural degradation of the chlorinated compounds by reductive dehalogenation. Normally, when we investigate a site that has been contaminated with chlorinated compounds such as trichloroethylene or tetrachloroethylene, we find breakdown products occurring at the site. The breakdown products usually have progressively less chlorine substitutions. These reactions were summarized in Figure 4.

Based on these natural reactions, researchers had been encouraged to find that bacteria were able to degrade these compounds. While several bacteria were found that could degrade chlorinated hydrocarbons, the research has had very limited success when trying to switch to field conditions. Work has now switched to the concept that the environment in the aquifer encourages the development of a diverse biological community that can be responsible for the dehalogenation. Recent work has concentrated on methods to change the environment to conditions that will enhance microorganisms that are capable of producing the desired reactions.

We now recognize that enhancements in these situations will be centered upon compounds that will help to create the required environmental conditions. For example, the easiest method to create highly reduced conditions in the aquifer is to add simple sugars. Bacteria using these organics will deplete all oxygen and nitrate available and lower the redox potential where dehalogenation reactions occur. This work is just beginning. The reader will have to continue to review the literature for articles updating refinement of methods of changing the environment to encourage the desired reactions. Several authors have written about natural conditions that facilitate biological dehalogenation (Hinchee et al., 1994; Semprini et al., 1992).

The other method that has been applied for the *in situ* destruction of halogenated compounds has been the use of a cometabolite. The two main cometabolites have been methane and the family of benzene ring compounds. Work at Stanford (Semprini et al., 1992) and by others has shown that there are several bacteria that can use methane as a cometabolite for the degradation of chlorinated

Table 9   Cost of Oxygen for *In Situ* Treatment

Oxygen-supply method	System flow rate	Estimated treatment time (days)	Percent of site treated	Cost capital ($)	Operation and maintenance cost ($/month)	Total cost ($)	Contaminant treatment cost ($/lb)
Air sparging	15 wells @ 2 cfm/well	1716	41	35,000	2000	148,000	90.3
Water injection	70 gpm	1580	85	77,000	2200	191,000	100.2
Venting (vapor control)	160 cfm	132	72	88,500	2500	99,000	13.4
Hydrogen peroxide	70 gpm	330	95	60,000	11,500	185,000	65.1
Nitrate injection	70 gpm	335	85	120,000	7500	203,000	77.2

Source: Adapted from Groundwater Technology, Inc. *Hazardous Waste Consultant* (July/August 1990).

Note: gpm = gallons per minute; cfm = cubic feet per minute.

hydrocarbons. These reactions occur under aerobic environments. Therefore, oxygen and methane must both be delivered to the contaminated area. Large pilot-scale tests are currently being performed, but at the writing of this book, no full-scale remediation has been completed.

The family of benzene-ring compounds has also been shown to serve as an excellent cometabolite for biodegradation of chlorinated hydrocarbons. Several *Pseudomonas* species produce enzymes capable of degrading chlorinated hydro-carbons while degrading benzene, toluene, benzoic acid, and other simple ring compounds. This work has been successful, and reports have shown preliminary success in field trials. The main problem with the introduction of ring compounds into the environment is the inherent hazardous nature of the ring compounds themselves. While cleanup of the chlorinated hydrocarbons is desirable, it is normally difficult to convince a regulator to approve the injection of a hazardous compound into the aquifer.

Delivery of Required Enhancements

The final area that must be discussed is the delivery of the enhancements to the aquifer and vadose zone. As discussed in Chapter 1, the delivery will be dependent upon the chemical properties of the compound in relationship to the carrier and to the geological limitations of the site. As can be seen in the above discussion, all of the enhancement methods require delivery of some type of compound. In Chapter 1, we discussed the two carriers available for underground situations, water and air. The oxygen transfer section discussed the advantages of water and air as a method to carry oxygen into the contaminated zone. Air has a tremendous advantage as a carrier due to the high content of the oxygen in the air and the high movement rate of air through most unconsolidated geological formations. Some of the organic compounds will have to rely on water due to low volatility, but high solubility.

The work with methane has been able to use air (gaseous) as the carrier, while the ring compound applications have used water as the carrier. Delivery of most inorganic material is by water. Most inorganics have good solubility in their ionic forms.

The other group of enhancements that may be required at a site are nutrients. The main nutrients are nitrogen and phosphorous in the form of ammonia or nitrate (for certain environments) and phosphate. Nutrient addition in the form of ammonia and phosphate are not limited by solubility. However, it is not a good idea to place too many nutrients in the acquifer at one time. Both organic contaminant and nutrient should be used up at the end of an *in situ* cleanup of an aquifer. There are several reasons for this. High concentrations of ammonia can change the pH or may be directly toxic to the bacteria. Also, excess ammonia can create its own oxygen demand and create nitrates. There is the chance of secondary plumes of nutirents.

The main thing to remember is that the geology controls the delivery of these materials. When *in situ* bioremediation fails, it is usually due to a lack of under-standing of the geological conditions at the site. Intrinsic bioremediation has shown us that most of the bacteria and subsequent reactions occur naturally at a site. Enhancement relies on the presence of these naturally occurring microorganisms. The *in situ* design solves the practical problems of delivery.

SUMMARY

Biochemical reactions are an important part of almost all remediations. The investigator must understand the potential bacterial affects at the site in order to correctly plan and interpret the remedial investigation. The remedial designer must include biochemical reactions in order to develop the most cost-effective design. When microorganisms are used as the basis of a remediation, the designer must include all of the intrinsic reactions that are ongoing at the site. Enhancements to these reactions will be limited by the delivery of the required material. *In situ* biochemical reactions are very powerful tools for the cleanup of contaminated sites. The reader must incorporate all the above ideas into a successful use of this technology.

REFERENCES

Baedecker, M.J., Siegel, D.I., Bennett, P.E., and Cozzarelli, I.M. The fate and effects of crude oil in a shallow aquifer: In the distribution of chemical species and geochemical facies: U.S. Geological Survey Water-Resources Investigations Report 88-4220. Reston, VA: U.S. Geological Survey, 1989.

Barbaro, J.R., Barker, J.F., Lemon, L.A., Gillham, R.W., and Mayfield, C.I. "In-situ cleanup of benzene in groundwater by employing denitrifying bacteria." Prepared for the Canadian Petroleum Products Institute, Ottawa, Ontario and Environment Canada, Wastewater Technology Centre, Burlington, Ontario, CN, CPPI Report No. 91-6, 1991.

Barrio-Lage, G., Parsons, F.Z., Nassar, R.J., and Lorenzo, P.A. "Sequential Dehalogenation of Chlorinated Ethenes." *Environ. Sci. Technol.,* vol. 20, pp. 96–99, 1986.

Bergey's Manual of Determinative Bacteriology, 8th ed., Buchanan, R.E. and Gibbons, N.E., Eds., Baltimore, MD: Williams & Wilkins, 1974.

Bouwer, H. *Groundwater Hydrology,* New York: McGraw-Hill, 1978.

Brock, T.D. *Biology of Microorganisms.* Englewood Cliffs, NJ: Prentice-Hall, 1979.

Bull, A.T. *Contemporary Microbial Ecology,* D.C. Ellwood, J.N. Hedger, M.J. Lathane, J.M. Lynch, and J.H. Slater, Eds., London: Academic Press, 1980.

Bumpus, J.A., Tien, M., Wright D., and Aust, S.D. "Oxidation of Persistent Environmental Pollutants by a White Rot Fungus." *Science,* vol. 228, pp. 1434–1436, June 1985.

Cooper, W.J. et al. "Abiotic Transformations of Halogenated Organics. 1. Elimination Reaction of 1,1,2,2-Tetrachloroethane and Formation of Chloroethene." *Environ. Sci. Technol.,* vol. 21, pp. 1112–1114, 1987.

Dragun, J. "Microbial Degradation of Petroleum Products in Soil," *Proc. Conference on Environmental and Public Health Effects of Soils Contaminated with Petroleum Products,* October 30-31, 1985, University of Massachusetts, New York: John Wiley & Sons, 1988a.

Dragun, J. *The Chemistry of Hazardous Materials,* Silver Springs, MD: The Hazardous Materials Control Research Institute, 1988b.

EPA, "Compilation of Groundwater Models." EPA/600/R-93/118, U.S. Environmental Protection Agency, Office of Research and Development, Washington, D.C., May 1993.

Grady, C.P. "Biodegradation: Its Measurement and Microbial Basis." *Biotechnol. Bioeng.,* vol. 27, pp. 660–674, 1985.

Hinchee, R.E., Anderson, D.B., Metting, Jr., F., Blaine, S., and Gregory, D., *Applied Biotechnology for Site Remediation,* Chelsea, MI: Lewis Publishers, 1994.

Howard, P.H. et al. *Environmental Degradation Rates,* Chelsea, MI: Lewis Publishers, 1991.

Kleopfer, R.D. et al. "Anaerobic Degradation of Trichloroethylene in Soil," *Environ. Sci. Technol.,* vol. 19, pp. 277–280, 1985.

Kobayashi, H. and Rittman, B.E. "Microbial Removal of Hazardous Organic Compounds." *Environ. Sci. Technol.,* vol. 16, pp. 170A–183A, 1982.

Lovley, D.R., Baedecker, M.J., Lonergan, D.J., Cozzarelli, I.M., Phillips, E.J.P., and Siegel, D.J. "Oxidation of aromatic contaminants coupled to microbial iron oxidation." Nature, 339: 297–300, 1989.

McCarty, P.L. "Anaerobic Biotransformation of Chlorinated Solvents (Abstract)," in *Biological Approaches to Aquifer Restoration, Recent Advances and New Opportunities,* CA: Stanford University, 1986.

Moat, A.G. *Microbial Physiology,* New York: Wiley-Interscience, 1979.

Nyer, E.K., *Groundwater Treatment Technology,* 2nd ed., New York: D Van Nostrand-Reinhold, 1992.

Norris, R.D., Hinchee, R.E., Brown, R., McCarty, P.L., Semprini, L., Wilson, J.T., Kampbell, D.H., Reinhard, M., Bouwer, E.J., Borden, R.C., Vogel, T.V., Thomas, J.M., and Ward, C.H., *In Situ Bioremediation of Ground Water and Geological Material: A Review of Technologies,* EPA/600/R-93/124, Robert S. Kerr Environmental Research Laboratory, Office of Research and Development, U.S. Environmental Protection Agency, Ada, OK, July 1993.

Parsons, F., Wood, P.R., and DeMarco, J. "Transformations of Tetrachloroethane and Trichloroethane in Microcosms and Groundwater," *AWWA J.,* vol. 76, pp. 56–59, 1984.

Paul, E.A. and Clark, F.E. *Soil Microbiology and Biochemistry,* San Diego: Academic Press, 1989.

Rittman, B.E., Smets, B.F., and Stahl, D.A. "The Role of Genes in Biological Processes, Part V." *Environ. Sci. Technol.,* vol. 24, no. 1, pp. 23–29, 1990.

Rittmann, B.E. et al. "Principles of bioremediation," In *In-situ Bioremediation When Does It Work?* Committee on *In-Situ* Bioremediation, Water Science and Technology Board, Commission on Engineering and Technical Systems, National Research Council, Washington, D.C.: National Academy Press, pp. 23, 1993.

Semprini, L., McCarty, P., and Roberts, P. *Methodologies for Evaluating In Situ Bioremediation of Chlorinated Solvents,* CA: Stanford University, U.S. Dept. of Commerce, National Technical Information Service (NTIS), March 1992.

Stanier, R.Y., Adelberg, E.A., and Ingrahm, J.L. *The Microbial World,* Englewood Cliffs, N.J: Prentice-Hall, 1976.

Vogel, T.M. and McCarty, P.L. "Transformations of Halogenated Aliphatic Compounds." *Environ. Sci. Technol.,* vol. 21, no. 8, 1987a.

Vogel, T.M. and McCarty, P.L. "Abiotic and Biotic Transformations of 1,1,1-Trichloroethane Under Methanogenic Conditions," *Environ. Sci. Technol.,* vol. 12, pp. 1208–1213, 1987b.

Wiedemeir, T.H., Downey, D.C., Wilson, J.T., Kampbell, D.H., Kerr, R.S., Miller, R.N., and Hansen, J.E. "Technical Protocol for Implementing the Intrinsic Remediation with Long-Term Monitoring Option for Natural Attenuation of Dissolved-Phase Fuel Contamination in Ground Water (draft)." Air Force Center for Environmental Excellence, Brooks Air Force Base, San Antonio, TX, August 29, 1994.

Wilson, J.T., Pfeffer, F.M., Weaver, J.W., Kampbell, D.H., Wiedemeier, T.H., Hansen, J.E., and Miller, R.N. "Intrinsic bioremediation of JP-4 jet fuel." Proc. Symposium on Intrinsic Bioremediation of Ground Water. Denver, CO. August 30 - September 1, 1994.

Wood, P.R., Lang, R.F., and Payan, I.L. *Groundwater Quality,* C. H. Ward, I.S. Giger, and P.L. McCarty, Eds., New York: John Wiley & Sons, 1985.

4   VAPOR EXTRACTION AND BIOVENTING

Sami Fam

INTRODUCTION

The vapor extraction and bioventing technologies induce air flow in the subsurface using an above-ground vacuum blower/pump system. Adequate air movement within the contaminated zones is of primary importance to the success of the vapor extraction system (VES). The induced air flow brings "clean" air in contact with the contaminated soil, non-aqueous phase liquid (NAPL), and soil moisture. The contaminated soil gas is drawn off by the VES and the air in the soil matrix becomes recharged with new vapor phase contamination as the soil/pore water/soil gas/NAPL partitioning is re-established.

Bioventing, or bioenhanced vapor extraction, is a remedial method similar to vapor extraction in that it relies upon an increase in the flow of air through the vadose zone. Vapor extraction is performed to volatilize the hydrocarbon constituents *in situ*. In bioventing, the increase in the flow of air provides oxygen in the subsurface to optimize natural aerobic biodegradation, which becomes the dominant remedial process. While the design criteria for vapor extraction and bioventing are different, once the physical system is in operation both processes occur. Compounds that are volatile move with the air, and compounds that are degradable have an increased rate of degradation. The design sections of this chapter will show how to set up the process so that one process is favored over the other.

Figure 1 shows the basic components of a vapor extraction system. Subsurface vapors are withdrawn through an extraction well that may be vertically or laterally constructed. Recovered vapors are routed to an above-ground vapor treatment unit, if required. The key to a successful design is to place the wells and equipment so that when the system is in operation an air flow pattern is created across the entire section of the unsaturated zone that is contaminated. The designer must also be careful that the air does not move through a small percentage of the area due to porous geology (short circuiting).

0-87371-995-6/96/$0.00+$.50
© 1996 by CRC Press, Inc.

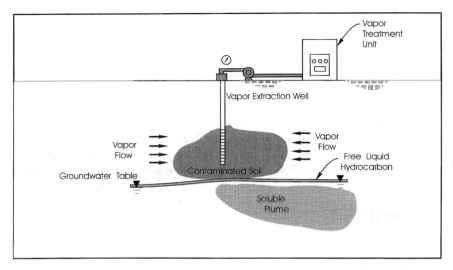

Figure 1   Basic VES system.

The distribution of contaminants within the four phases (pure contaminant, vapor phase, soil moisture, and adsorbed to soil) can be represented by mathematical equations, and the distribution can be simulated by computer models. For large projects or systems where accurate prediction of system performance is required, pilot testing of the VES is often performed. The design engineer must gather information during the pilot testing activities or the predesign phase of the project to calibrate the simulation models which will in turn aid in ascertaining achievable closure criteria (to be compared with risk assessment goals), closure time-frames, and optimal system operating parameters (vacuum, flow rates, moisture content, well location, well screening, number of wells). Projects for which this level of sophistication is conducted may include litigation-oriented SUPER-FUND projects or projects where the site owner desires accurate prediction of the lifetime of the remedial system.

For small projects, such as a gasoline service station with shallow vadose contamination in a high-permeability geology, modeling often is not conducted. If simulation models are not utilized, the design engineer must choose the required subsurface air flow velocities to achieve cleanup goals using published literature values or other empirical design concepts. At a VES typical of a service station, the engineer is often constrained by subsurface structures for well locations. Due to these constraints, the systems may not be optimally designed and are often overdesigned to overcome site access limitations.

A vacuum pump or blower is the tool that is used to create subsurface air flow (Figure 1). The vacuum created at the extraction well head is an indication of the subsurface soil resistance to air flow. If the subsurface is very porous (sands/gravel) there will be very little vacuum at the extraction well head regardless of vacuum pump that is utilized (i.e., low vacuum-capability regenerative blower or high vacuum liquid-ring type pump). If there is little resistance, there will be little resultant vacuum. If one increases the flow from a given well, more vacuum application will be required because of increased subsurface resistance created by the higher flow. Vacuum is an indication of subsurface resistance to flow and is not the variable that is critical to VES success. Sufficient air flow is the critical variable. The vacuum application is a system operational parameter that allows creation of the desired air flow.

This chapter will discuss the theory of vapor extraction, site and chemical parameters that are used to predict its applicability, modeling of VES, pilot testing, system design criteria, and the biological enhancement that results from vapor flow (bioventing); vapor phase treatment options will be discussed in Chapter 7.

CONTAMINANT PARTITIONING IN THE SUBSURFACE

Contaminants that are released to the environment will be distributed in the subsurface in a manner consistent with their physical properties. This subsurface distribution (in pore water, vapor, adsorbed to soil, or in pure NAPL) is termed "partitioning". Partitioning is related to properties of the soil (type, moisture extent, etc.), as well as the contaminants (vapor pressure, solubility, etc.).

Transfer of the contaminants between phases (vapor, dissolved, free phase, and adsorbed) is affected by the relative affinity of the contaminant to each phase. These affinities can be evaluated using the constituent partitioning coefficients to the various phases. The interphase partitioning coefficients can be expressed as the concentration ratio of the constituents in each phase, and this ratio is dictated by the equilibrium relations in the subsurface (Equations 1 to 3). Since, to a large extent, the interphase transfer is governed by these equilibrium partitioning relationships, the most effective remediation will create the subsurface conditions that will drive the interphase transfer towards the phase(s) that allow for the most efficient mass removal. Site remediation, therefore, can be viewed as implementing changes or perturbations to the subsurface that will drive chemical and biological processes toward the site remediation goals (Sims, 1990). The subsurface change that is affected during vapor extraction is replenishment of the subsurface soil vapor (air is the carrier), therefore driving the contamination to the vapor phase where it is collected for above-ground treatment. Figure 2 is a schematic of the environmental compartment model showing the goals of vapor extraction.

When a volatile NAPL is present on the soil, the bulk of the mass removal by VES will come from direct volatilization of the NAPL. This would be similar to a fan blowing past a pool of gasoline. Research workers (Hoag et al., 1989) have shown that the bulk (over 95%) of the NAPL can be removed within passage of several hundred pore volumes of air through the experimental soil columns (NAPL within the dry soil void space). In field applications, where air flow is usually over the NAPL layer rather than through it, VES still often recovers the bulk of the NAPL within several hundred pore volumes. Under conditions of NAPL presence, mass removal rates are often linearly correlated with air flow

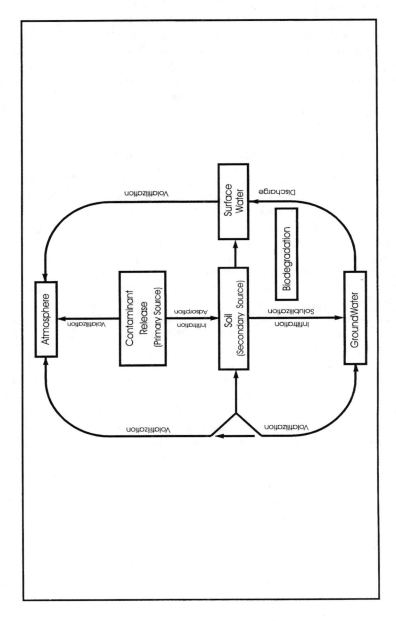

Figure 2   Environmental compartment model for VES.

rates. When NAPL is not present in the subsurface, air flow requirements become very different and are often governed by nonequilibrium rate-limiting conditions.

The following section describes the partitioning of contaminants in the subsurface. These equations are the basis of numerical simulation models that attempt to predict the remediation process of vapor extraction. Models achieve this prediction by repeatedly evaluating the "new" partitioning relationships after passage of "clean" air past the contaminated soil. Modeling is discussed later in this chapter.

Under moist soil conditions, contaminant partitioning in the vadose zone (no residual NAPL) can be described by the following equation (1). Figure 3 is an illustration of the equation:

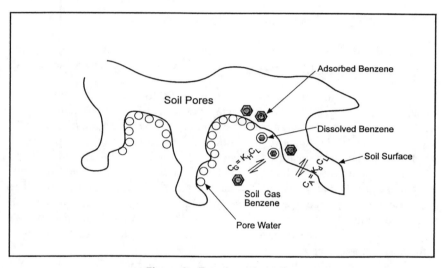

Figure 3 Equation schematic.

$$C_T = p_b \, C_A + \theta_L \, C_L + \theta_G \, C_G \tag{1}$$

where

C_T = total quantity of chemical per unit soil volume,
C_A = adsorbed chemical concentration,
C_L = dissolved chemical concentration,
C_G = vapor concentration,
p_b = soil bulk density,
θ_L = volumetric water content,
θ_G = volumetric air content.

The equilibrium relationship between vapor concentration (C_G) and the associated pore water concentration (C_L) is given by Henry's Law:

$$C_G = K_H \, C_L \tag{2}$$

where K_H = Henry's Law constant.

Henry's Law is often thought of in relation to air stripping. In air stripping, air removes dissolved volatile organic chemicals (VOCs) from the water stream. The efficiency of this removal process is related to the Henry's Law constant of the compound. Under moist soil conditions, the extracted vapors during vapor extraction similarly remove VOCs from water. The success of this removal process is similarly related to the Henry's Law constant of the compound.

Likewise, the relationship between equilibrium solution concentration and adsorbed concentration is given by:

$$C_A = K_d \, C_L \tag{3}$$

where K_d (L^3/M) is the distribution coefficient expressed as $K_d = f_{oc} K_{oc}$, where f_{oc} is the mass fraction of organic carbon, and K_{oc} is the organic carbon partitioning coefficient.

Equation 3 is generally considered to be valid for soils with high organic content ($f_{oc} > 0.1\%$ solids). For soils with lower organic carbon content ($f_{oc} < 0.1\%$), sorption to mineral grains may be dominant (Piwoni and Bannerjee, 1989; Brusseau et al., 1991). Adsorption is further discussed later in the chapter.

Mathematical relationships can also be developed to quantify biological and or other reaction transformations of the chemical contaminants. These relationships subsequently can be used to develop mathematical models of the subsurface conditions under advective air movement conditions. Mathematical models can be used to simulate subsurface changes caused by the VES (perturbation) and allow the user to select the most efficient perturbation that leads to the most contaminant mass removal.

The above equations have defined partitioning of contaminants in the subsurface without air movement. Partitioning without advective air movement occurs via diffusion. The VES induces air flow (advective) past the contaminated zone; therefore, under VES operating conditions, both diffusive and advective transport are occurring.

Under the assumptions of moisture distribution across the soil and incompressible air phase, the advective-dispersive transport equation in Cartesian coordinates can be written as (Armstrong et al., 1994; Gierke et al., 1989):

$$L(C_G) = \theta_G D_{ij} \frac{\partial}{\partial x_i} \frac{(\partial C_G)}{(\partial x_j)} - \theta_G v_i \frac{\partial C_G}{\partial x_i} = \theta_G \frac{\partial C_G}{\partial t} + \theta_L \partial \frac{C_L}{\partial t} + P_b \frac{\partial C_A}{\partial t} \tag{4}$$

$$i, j = x, z$$

where subscripts G, L, and A designate the gaseous, dissolved, and sorbed phases of the contaminant. C_A, C_L, C_G, and P_b, are as defined above. θ_G and θ_L are the volumetric gas and water contents. The air velocity component is derived from the air continuity equation and the subsequent application of the Darcy equation to pressure. The continuity equation states that the same mass of material entering a unit volume of space must also exit that volume space (without biodegradability) The Darcy equation relates groundwater velocity to hydraulic gradient. In this application, the Darcy equation $\left(V = \dfrac{k}{\phi} \dfrac{dh}{dl} \right)$ is applied for air movement rather than ground water. The term D_{ij} is the dispersion tensor, defined in terms of longitudinal and transverse dispersivity and the diffusion coefficient. Equations 1, 2, and 3 can be substituted into Equation 4 in order to represent the transport equation in terms of the gaseous phase only to yield:

$$K_d = f_{oc} K_{oc} \tag{5}$$

$$\theta_G D_{ij} \frac{\partial}{\partial x_i} \frac{(\partial C_G)}{(\partial x_j)} - \theta_G v_i \frac{\partial C_G}{\partial x_i} = R\theta_G \frac{\partial C_G}{\partial_t} \tag{6}$$

$$R = 1 + \frac{\theta_L}{\theta_G H} + \frac{P_b K_d}{\theta_G H}$$

Equations 5 and 6, therefore, are a mathematical representation of partitioning under diffusive and advective conditions. Actual field observations however, indicate that the equations are only valid for diffusion-dominated or weakly advective conditions.

Under strongly advective conditions that can be found while operating VES, the above equations do not account for the long tailing effect that is observed in field applications. Tailing is the phenomenon that is often observed in VES applications whereby the contaminant mass removal rates are slower, and the residual mass of contaminant adsorbed to the soil after vapor extraction is greater than what would be predicted by the equilibrium equation, Equation 6. Equilibrium predicted nontailing- and nonequilibrium-type tailing effects are shown schematically in Figure 4.

As previously discussed in Chapter 2, the performance of a system in removing contaminants changes with time. In most VES applications, the initial stages of the project yield the highest mass removal. The tailing effect implies that efficiency is decreasing with system lifetime and that closure goals or system operational modifications should account for this temporal change.

Several authors (Armstrong et al., 1994; Gierke et al., 1989; Sleep and Syikey, 1989; Brusseau et al., 1991) have shown that the tailing effects can be represented by first order mass transfer, nonequilibrium, physical, and/or chemical processes. These nonequilibrium modifications to the equilibrium relationships (Equations 1 to 6) will not be presented in this chapter; however, the reader is referenced to the appropriate research publications for details. In brief, the nonequilibrium notions relate to the existence of a rate-limiting criteria governing the mass transfer process for VES. Under fully wetted soil conditions without NAPL, the

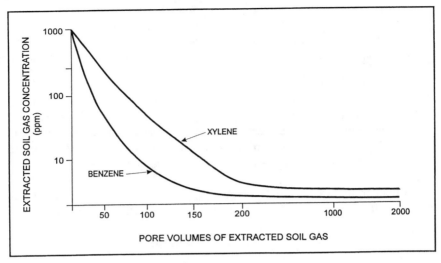

Figure 4 Equilibrium assymptotic tailing effect.

rate-limiting step may be transferred across the air/water interphase, the soil/water interphase, dead-end micropore effects, and/or a combination of all effects. Simply stated, these nonequilibrium effects slow down the vapor extraction process once the bulk of the contamination has been removed. Armstrong et. al. (1994) have conducted a sensitivity analysis on several of these nonequilibrium rate limiting conditions.

A sensitivity analysis or demonstration of the physical limitation of the VES ability is a powerful tool in ascertaining the capability of the system and hence achievable closure goals. This can be utilized to negotiate reasonable closure criteria with regulatory agencies or to modify system operation during the lifetime of the project to minimize expenses while maximizing mass removal. Lastly, if the VES limitations are known, alternate technologies after the VES lifetime may be considered if required.

AIR FLOW REQUIREMENTS AND CAPABILITIES

The need to understand and predict the subsurface mass transfer relationships relates to the practical need to deliver the required air flow to achieve the remedial goals. Often the designer will only want the minimum subsurface air movement to achieve the remedial goals, since excessive air flow results in larger, expensive off-gas treatment equipment as well as higher operating costs.

In instances where NAPL is present in pockets, pools, or as a layer atop the ground water, mass removal often will be linear or semilinearly rated to the air flow. This does not imply that if NAPL is present, high air flow is required, since often the NAPL is removed rapidly, leading to site conditions that may not require further high air flow. Air flow generation capability (air flow that can be generated based upon subsurface soil conditions) and the air flow requirement (to achieve remediation) must be met in order to appropriately install a VES.

Air Flow Capability

Figure 5 presents predicted air flow rates per unit well screen (from an extraction well) depth for a 4-in. diameter extraction well and a wide range of soil permeabilities and applied vacuums (Johnson et al., 1990; Johnson et al., 1991). The graph was generated by solving the logarithmic Jacob's equation and assuming a 40 ft radius of air collection. The Jacob's equation is the same Jacob's equation used in groundwater applications to determine the zone of influence of a pumping test. In this instance, however, it is utilized for subsurface air flow rather than groundwater flow. This figure provides an excellent screening tool to determine the necessary vacuum equipment to generate the required air flow. This will be further discussed later in the chapter.

The measurement of vacuum at locations away from the vapor extraction well implies that subsurface air flow is present at that location. Subsurface vacuum is easily to measured in the field; therefore, it is often measured in place of subsurface air flow during VES pilot testing and application.

Figure 6 illustrates how quickly vacuum measurement profiles "die off" away from the extraction well point. This quick decay of induced vacuum readings in turn implies that subsurface air flow quickly "dies off" at increasing distances from the extraction well.

Air Flow Requirements

Delivering the required air flow to achieve the cleanup criteria is the basic design goal for VES installation. This basic design goal, however, remains the most difficult to predict due to our limited understanding of subsurface conditions. This understanding is required to quantify the mathematical relationships used in formulating the simulation models. The most distant location from the extraction well should receive sufficient air flow to achieve remediation. The most distant location from the extraction well is termed the "zone of influence". Sufficient wells are spaced in the contaminated area to deliver the minimally acceptable air flows across the entire site.

The required air flow at the most distant location can be determined by (1) comparison of published literature values for similar soil types and contaminants, (2) by conducting bench scale column tests to determine this required air flow for the contaminants and soil conditions (discussed later in the chapter), or (3) by conducting computer simulation modeling (discussed later in the chapter).

EVALUATION OF CONDITIONS WHERE VES IS APPLICABLE

Vapor extraction system efficiency is affected by parameters relating to the contaminants to be removed and by variables relating to the site to be remediated.

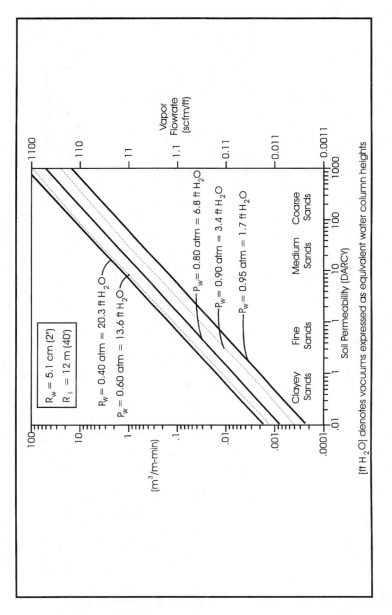

Figure 5 Air flow generation plot.

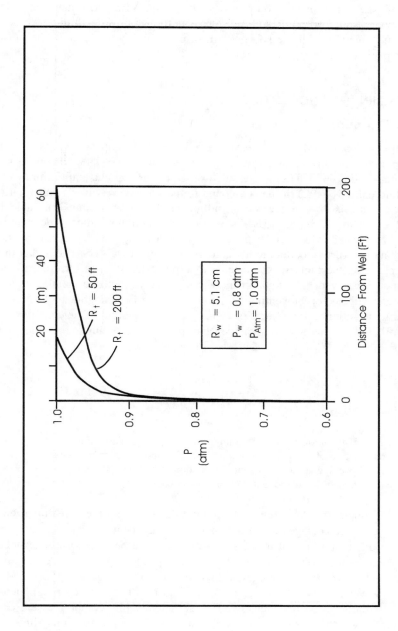

Figure 6 Vacuum tailing effect.

Contaminant properties that affect VES are vapor pressure, solubility, Henry's Law constant, biodegradability, and other molecular structure properties (Jury et al., 1990). Vapor pressure tends to be the most important chemical parameter affecting VES performance. Soil properties that affect VES performance include soil porosity, soil adsorption, soil moisture, site topography, depth to water, and site homogeneity. Of all the soil properties, permeability tends to be the most critical parameter relating to system success.

CONTAMINANT PROPERTIES

Vapor Pressure

Vapor pressure is the parameter that can be used to estimate the tendency of a compound to volatilize and partition into the gaseous state. The vapor pressure of a compound is defined as the pressure exerted by the vapor at equilibrium with the liquid phase (NAPL) of the compound in the system at a given temperature. Vapor pressure is often expressed in millimeters of mercury (mmHg). Table 1 provides a listing of vapor pressures of some common environmental contaminants. When chemicals exist in pure form (NAPL), the vapor pressure of the contaminant is very important to its removal efficiency by a VES. The higher the vapor pressure, the more it is amenable to vapor extraction. In turn, the lower the vapor pressure of the contaminant, the less likely it will be volatilized and the greater reliance on bioventing (if the contaminant is biodegradable) for successful remediation. For mixtures of compounds (i.e., gasoline), the composition of the mixture also has a bearing on the vapor pressure according to the following relationship:

$$P_i{}^* = X_i\,A_i\,P_i{}^\circ \tag{7}$$

where

$P_i{}^* =$ equilibrium partial pressure of component i in the organic mixture,
$X_i\ \ =$ mole fraction of component i in the organic compound mixture,
$A_i\ \ =$ activity coefficient of component i in the organic compound mixture,
$P_i{}^\circ =$ vapor pressure of component i as a pure compound.

For example, the above relationship states that the vapor pressure of benzene is related to the percentage (mole fraction) of the component, benzene, in gasoline.

If NAPL is not present in the soil, vapor pressure becomes a less accurate predictor of VES efficiency since other relationships (adsorption to soil, moisture content, etc.) become more important in governing system success. It should be noted however that even under conditions of no NAPL presence, a compound must be volatile to be removed by VES. Sufficiently high vapor pressure can therefore be viewed as a pre-requisite for successful vapor extraction. Although

Table 1   Contaminant Properties —
Vapor Pressure Parameters

Compound	Vapor pressure (mmHg)
Acetone	89 at 5°C
Benzene	76 at 20°C
Toluene	10 at 6.4°C
Vinyl chloride	240 at –40°C
o-xylene	5 at 20°C
Ethylbenzene	7 at 20°C
Methylene chloride	349 at 20°C
Methyl ethyl ketone	77.5 at 20°C
Trichloroethylene	20 at 0°C
Tetrachloroethylene	14 at 20°C

the definition of sufficiently high vapor pressure is rather subjective, 1 to 2 mmHg should be used as a guideline. Compounds with lower vapor pressures will likely be removed more slowly and greater reliance will be required on *in situ* biological breakdown of the compounds (bioventing).

Solubility

Aqueous solubility is one of the most important parameters governing the partitioning, transport, fate, and, therefore, the ultimate remediation of site contaminants. Solubility can be defined as the maximum amount of a constituent that will dissolve in pure water at a specified temperature. For organic mixtures (such as gasoline), solubility is additionally a function of the mole fraction of each individual constituent in the mixture according to Equation 8. Chapter 1 contains a table listing of pure water solubilities for some common environmental contaminants.

$$C_i^* = X_i \, A_i \, C_i^\circ \qquad (8)$$

where

C_i^* = equilibrium concentration of component i in the organic mixture,
X_i = mole fraction of component i in the organic compound mixture,
A_i = activity coefficient of component i in the organic compound mixture,
C_i° = equilibrium solute concentration of component i as a pure compound.

Under most vapor extraction scenarios, the vadose soil is relatively moist (10 to 14% by weight) and contaminants are generally dissolved in the soil pore water. Solubility is also a critical factor for the bioventing of contaminants, since biological degradation is enhanced or simplified if the contaminants are more available for microbial uptake by being dissolved in the pore water. A soil moisture of 12% by weight is generally required for adequate bioventing.

Henry's Law

The interaction of solubility and vapor pressure produces a behavioral modification that renders the additive effects of solubility and vapor pressure nonlinear. This interaction has particular impact on volatilization of organics from water. The reader is referenced to basic chemistry textbooks for the derivation of Henry's Law (Mahan, 1966). Henry's Law constant is functionally defined as the ratio of saturated vapor density to chemical solubility for a given compound:

$$H = C_{sg}/C_{si} \qquad (9)$$

where:

H = Henry's Law constant
C_{sg} = compound concentration in the vapor phase at the water/vapor interphase,
C_{si} = compound concentration in the water phase at the water/vapor interphase.

The Henry's Law relationship is depicted graphically in Figure 7. Under moist soil conditions, therefore, VES efficiency is Henry's Law dependent, much like VOC removal by air stripping. For example, although acetone is very volatile, it is not well remediated by a VES due to its high water solubility. Acetone tends to biodegrade readily, however, and is therefore amenable to bioventing.

In a similar manner that air-stripping efficiency is temperature dependent, VES efficiency has a temperature dependence. The temperature relationship is more complex for VES due to the existence of multiple system variables such as biodegradation, adsorption, etc. In general, higher temperatures in the vadose zone enhance volatilization, which improves operation of a VES. Increased temperature also enhances biodegradation, increases the rate of desorption, and weakens the adsorption binding.

Other Molecular Properties

There are several other molecular properties of the contaminant that influence the success of vapor extraction or bioventing. Although these properties are not as significant as vapor pressure, solubility, and Henry's Law and hence may be considered secondary, these secondary properties often may be the rate-limiting criteria to site remediation. Compound size, molecular weight, electronegativity, and polarity affect adsorption of the contaminant to soil particles and its travel through soil micropores. Larger, bulkier (branched chains) molecules travel more slowly within soil micropores and tend to adsorb more strongly to soil surfaces. Once the bulk of the more accessible (from large pores, and not directly adsorbed to soil) contaminant is removed, the "final molecules" removal tends to be rate limiting. Polarity and electronegativity relate to the effective charge of a compound and its interaction with the surface charge of the soil. These topics are further discussed below.

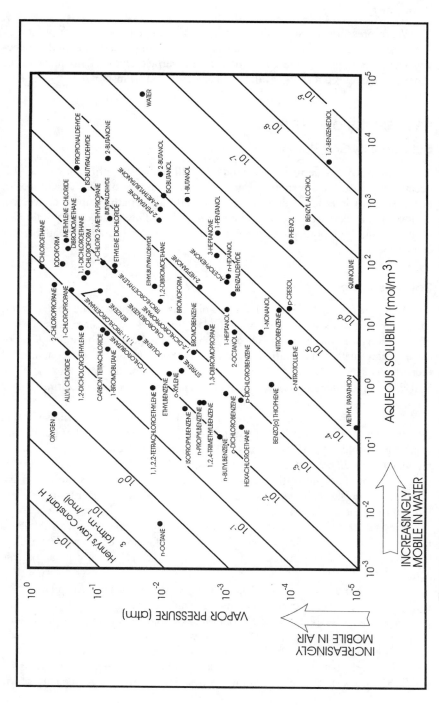

Figure 7 Henry's Law constant as a function of vapor pressure and solubility. (After Lyman et al., 1982).

Summary

Vapor pressure and solubility are the most dominant contaminant properties affecting the success of a VES. Secondary contaminant properties include molecular weight, compound size and structure, and surface charge. This presence of water in the soil reduces the available space for vapor transport.

PROPERTIES OF THE SOIL

Bulk Density/Soil Porosity

Decreasing soil permeability generally will reduce the efficiency of a VES because the diffusive transport from the soil matrix to the soil surface and in turn to the soil gas is reduced due to the increased path length (path to convective airflow) and a decrease in cross-sectional area for air flow. A secondary influence of decreasing soil porosity is the increase in soil surface areas available for contaminant binding.

Soil Adsorption

There are two methods by which the soil can adsorb the organic contaminant. The soil organic content and its mineral adsorptive surface sites are both capable of contaminant adsorption. The adsorption of contaminants to the soil organic or mineral clay surfaces tends to increase the immobile fraction of the contaminant (adsorbed to the soil) and to decrease the vacuum extraction system efficiency. Soil adsorptive interactions become particularly important (rate limiting) under drier soil moisture conditions.

The clay soil composition is strongly correlated to many water transport and retention properties which influence the volatilization process. Clay holds onto water tightly and is poorly transmissive. This presence of water in the soil matrix reduces the available space for vapor transport. The effects of soil moisture or pore water on VES are further described below. Clayey soils also tend to have small micropores, making transport path lengths longer. The longer transport path lengths reduce the efficiency of vapor extraction.

Clay mineral surfaces tend to have a net negative charge; they will influence the adsorption reaction of most compounds in some fashion. Clays are an excellent adsorbent of positively charged actions (such as metals) or very polar organic actions.

Soil total organic carbon (TOC) matter content is strongly correlated with the binding capacity for organic chemicals. This general correlation is valid as a general guidance despite the wide variation in the makeup of soil organic content, its state of decomposition, and, consequently, its binding capacity. The soil adsorption correlation coefficient is likely to be highest when the soil organic content is high, but observations of strong binding have been documented for soil TOC as low as 0.1%. A relationship describing the binding capacity of the soil to organics was defined in Equation 3. The parameter K_{oc} in Equation 3 is defined as:

$$K_{oc} = \mu g \text{ of contaminant adsorbed/gram of soil TOC } \mu g/mL \text{ solution} \quad (10)$$

Due to the preponderance of other soil organic matter surfaces and the nonpolar nature of most organic contaminants, there usually is little correlation between clay content and VOC adsorption (Jury et al., 1990).

Most organic contaminants are more easily adsorbed to the soil than they are desorbed. It therefore takes much longer and requires more energy to remove the contaminants from the subsurface than it does to spill them. This phenomenon, known as "hysterisis", tends to slow down the vapor extraction process than would be predicted by simple adsorption isotherm data.

Soil Moisture

Soil moisture is a very important parameter for VES success. High soil moisture content limits air advection travel pathways by occupying void space. Since movement of VOCs is much faster in the gas phase than in the liquid phase, it would be expected that VOC removal by vacuum extraction would be enhanced by decreasing soil moisture. This trend is not always observed. The lack of soil moisture allows contaminant adsorption to soil surfaces to play a more prominent role in mass transfer as the water particles are removed from the surface. If the soil adsorptive capacities are strong, the benefits of soil dewatering (increased air travel pathways) may be partially offset by this increased soil binding capacity (Sims, 1990; Thibaud et al., 1993). The moisture content at which a decrease in vapor concentration (during VES operation) is often termed the "critical" moisture content and is empirically defined as one monolayer of water molecules coating the soil surface. Recent research observations hypothesize that water particles may act to "kick out" adsorbed organics, thereby enhancing VES operation under certain conditions. Figure 8 provides a schematic of this concept. If soil adsorptive capacities are very weak (low TOC), it may be advantageous to conduct vapor extraction under drier conditions.

The notion that an optimal moisture content exists for a given contaminant, based upon its Henry's Law constant and upon the soil binding capacity, should allow for some process control and optimization of the VES performance. Although theoretically possible, this notion is rarely applied in the industry due to limited modeling resources and incomplete understanding of the partitioning coefficients of the site.

Site Surface Topography

Site surface topography can greatly influence the success of a VES. Ideally the site should be covered by an impermeable surface such as pavement or concrete. The covered surface serves two functions. First, it minimizes the infiltration of

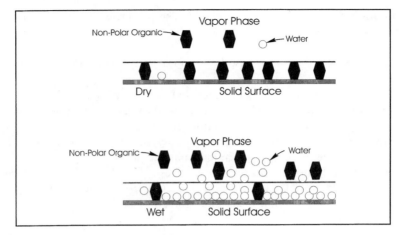

Figure 8   Effect of soil moisture on VOC adsorption.

rainwater to the vadose soils and consequently allows some control over soil moisture. Second, the covered surface eliminates the possibility of extraction well short-circuiting (Figure 9), where the majority of the extracted volume of air is coming from near the ground surface, and locations more distant from the extraction wells receive minimal air flow.

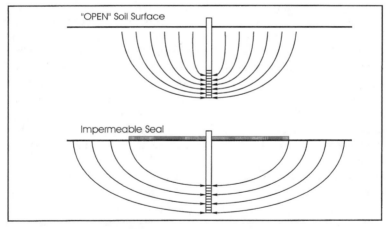

Figure 9   Effect of surface seal on vapor flowpath.

Short-circuiting also may be due to the presence of higher permeability zones such as utility trenches. If operation of a VES in a zone that is prone to short-circuiting is conducted, a higher number of VES wells generally will be required. This will in turn lead to higher air flows and higher capital costs for air treatment devices.

In order to minimize the effects of surface short circuiting, wells should not be screened near the surface (5 ft). In instances where a surface seal is not available, plastic sheeting can be applied (preferably buried under 1 ft of cover) to enhance system performance.

Depth to Water Table

If the VES well penetrates the water table (use of a converted monitoring well for vapor extraction) and vacuum is applied to the well, the water table within the well will rise by an amount equal to the level of applied vacuum. A 60-in. water column vacuum at the well head will therefore result in a water table rise of 5 ft (Figure 10). If there are only 5 ft of well screen above the water table, the water table rise may clog the available well screen. This scenario may be encountered in a situation where the water table is shallow or when the well is inappropriately designed. Horizontal wells can be utilized in shallow water table situations, thereby enlarging the available screen length and reducing well-head vacuum, thus minimizing water uplift.

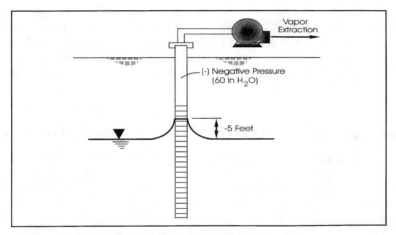

Figure 10   Effect of vapor extraction on water levels.

Installing the VES well with well screen above the water table will minimize water table uplift as the vacuum at the water surface will be less than at the well-head. As Figure 6 points out, well-head vacuum dissipates quickly away from the extraction well. As a general rule, the bottom of the VES well should be a minimum of 2 to 3 ft above the water table if possible to prevent this effect.

Site Homogeneity

Site homogeneity is very important to ensuring that air flow reaches all areas requiring remediation. The air carrier must flow past the contaminants if they are to be removed. Transport of the contaminants by the carrier air flow minimizes diffusion requirements for mass removal. This reduces the travel path length to remediation and expedites the cleanup. As an example, NAPL floating atop the water table takes longer to be volatilized because air flows over it rather than through it. Lab experiments where air is drawn through NAPL-saturated soils usually results in NAPL volatilization that is much faster than volatilization of NAPL floating on the water table.

Site nonhomogeneity can be partially alleviated by varying well screen designs to maximize air movement in contaminated zones, by fully opening some extraction wells (in low permeability zones) and closing others (in high permeability zones), and by possibly breaking the area into more than one zone based upon permeability. Wells in high permeability zones can be connected to a moderate vacuum blower, and wells screened in low permeability zones can be connected to high-vacuum, liquid-ring type pumps (Figure 11).

The presence of micro lenses of highly adsorptive, low permeability soils often will be rate limiting during VES operation (Figure 12). These highly adsorptive lenses are not accessed by the carrier air and rely on concentration gradient-driven diffusion.

At a VES site, the presence of utility trenches (generally constructed of high permeability fill material) or other high permeability air flow paths also may provide short-circuit pathways. In these instances, well screening or placement may require adjustment to accommodate the contaminant distribution. This adjustment usually requires deeper well screens and a higher density of extraction wells.

MODELING TOOLS FOR VAPOR EXTRACTION SYSTEM DESIGN

Computer simulation modeling is utilized to better design vapor extraction systems. This enhancement is due to the simulation of complex subsurface processes. This allows the model user to vary operating parameters and to observe a simulation of the result. For example, the model allows for comparing subsurface flow regimes with 10 or 20 extraction wells. The modeler can evaluate the benefits to be gained by installing the additional 10 wells and decide whether the added benefit is worthy of the additional installation and operational costs.

For the purposes of this discussion, three classes of modeling tools for use in VES design are described. This is a broad and crude model classification but is useful in presenting the basic requirements for VES design. The following three different types of models will be discussed:

1. Engineering design model
2. Air flow models
3. Multiphase transport models

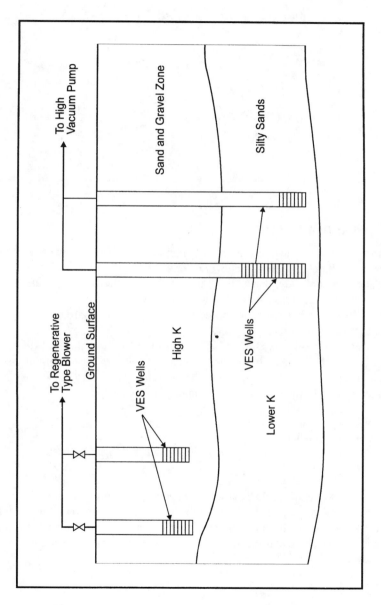

Figure 11 Two-zone venting system.

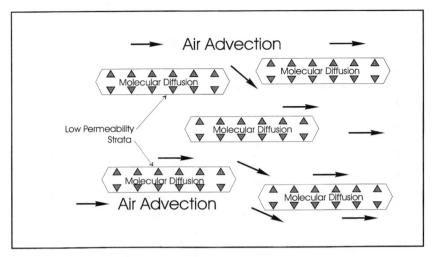

Figure 12   Advective/diffusive airflow schematic.

Engineering Design Model

Models are available that simulate vacuum losses in piping networks, vacuum equipment, well design, valving networks, and above-ground air treatment equipment. These models are particularly useful for the design of larger systems (greater than 10 extraction points) where these calculations may become cumbersome. These models often are based on solution of the Hardy Cross type of pneumatic flow equations.

One such model is the Dynkin VES design model (Geraghty & Miller, Inc., Andover, MA) that can assist in well placement, blower sizing, piping sizing, etc. The program operates in an interactive, iterative fashion to assist the user in finding an optimal system design. The input parameters are

1. Number of extraction wells
2. Gas flow equation coefficients (estimate based on pilot test)
3. Proposed system layout, including valves, length of pipe, diameters
4. Type of air-treatment device

The model calculates distribution of vacuum, pressure, and flow in all elements (pipes and wells) of the VES from the bottom of the vacuum extraction wells to the outlet of the air treatment device. The user can change the input parameters listed above and see how this changes the air flow and vacuum profiles in the system.

Flow Models

Flow-type models Airflow SVE (Waterloo Hydrogeologic Software, Waterloo, Ontario, Canada) HYPER VENTILATE (U.S. Environmental Protection

Agency), and AIR 3D (Geraghty & Miller, Inc., Millersville, MD) simulate air flow. They do not simulate mass transfer between the various compartments. These models are very simple to operate and allow the user to locate wells in order to achieve a predetermined vacuum or, conversely (AIRFLOW and AIR 3D), allow the user to translate vacuum influence readings into cross-sectional air flow at a given remote location. In either case the models allow for vacuum and air flow profile distributions across the generating two- or three-(AIR 3D) dimensional profiles.

Some of these models also show mass removal of NAPL (using vapor pressure and mass composition of the NAPL mixture) based upon a given flow configuration. The predictions however do not account for any mass removal beyond what is expected from NAPL venting.

These models do not account for volatilization from pore water, soil surfaces, or any equilibrium partitioning. If flow models are utilized, the user must select the minimally acceptable air flow across the site based upon published literature values for the contaminant of concern within similar soils. For example, if TCE is to be extracted from sandy soils, a minimally acceptable air flow across the most distant location from the extraction well must be selected. A very limited, but useful data base (minimally acceptable air flows) exists in the published literature for this purpose. Air flow models, therefore, allow for selection and optimization of well placement and well screening within the contaminated zone. This selection is based on a predetermined, minimal acceptable air flow.

Multiphase Transport Models

Multiphase transport modeling is usually not required for the majority of simple VES applications. Vapor extraction at a typical 1/4-acre, sandy soil, shallow water table service station does not require multiphase transport modeling. The installed systems are usually small and overdesigned and to a large extent rely on bioventing to supplement any shortcomings of vapor extraction system design, since the contaminants are biodegradable.

At more complex sites, however, where any or all of the following is desired, multiphase transport modeling can be very useful:

1. System layout and sizing optimization is desired due to the large size of the site (too expensive to overdesign).
2. System operating parameter optimization is desired.
3. Vapor extraction feasibility requires demonstration.
4. Contaminants may not be biodegradable.
5. Impacts to groundwater may be significant.
6. Closure time-frame prediction is required.

Solution of Equation 6 with the requisite modifications to account for non-equilibrium, nonlinear behavior is the basis for the multiphase transport models. Several models have been developed by researchers (Armstrong et al., 1994) but very few models that can adequately describe multiphase transport are commercially available.

Multiphase transport modeling allows for simulating the remediation process over time based upon the selected well layout. The multiphase transport models are often preceded by simple air flow models that locate extraction wells. The multiphase transport models simulate the VES perturbation (inducement of air flow) to the contaminant partitioning and predict contaminant concentrations within the various media at future points in time. The modeler can vary the soil moisture, air flow, or any other parameter of the site and observe the predicted effects.

A numerical model, therefore, is the best means of understanding the mass transfer between the liquid (soil pore water) and gas phase and the degradation of constituents into different species. MOTRANS, developed for the EPA by Parker and Kaluarachchi at The Virginia Polytechnic Institute and State University in Blacksburg, Virginia, is commercially available. MOTRANS can be used to simulate either two-phase flow of water and non-aqueous phase liquid (NAPL) in a system with gas present at a constant pressure, or three-phase flow of water, NAPL, and gas at variable pressure. Systems with no NAPL present or with immobile NAPL at a residual saturation may also be modeled by an option that enables elimination of the NAPL flow equation. The transport module can handle up to five components that partition among water, NAPL, gas, or solid phases assuming either a local equilibrium interphase mass transfer or first-order kinetically controlled mass transfer.

The flow of water and vapor, and the transport of constituents in the vadose zone, is a highly complex process. The equations governing these processes are strongly nonlinear, difficult to solve, and require extensive data input to characterize the physical properties of both the media and the fluids. In general, the principal limitation in applying modeling codes is characterization of the problem. Migration of constituents in the vadose zone is controlled by local heterogeneities, which may be difficult to define.

In addition, the physical properties characterizing the relative permeabilities and fluid retention characteristics are rarely collected. Multiphase flow and multicomponent transport require specification of permeability/saturation/capillary pressure relationships, air/water capillary retention function parameters, NAPL surface tension and interfacial tension with water, NAPL viscosity, NAPL density, maximum residual NAPL saturation, soil permeabilities and dispersivities, initial phase concentrations, equilibrium partition coefficients, component densities, diffusion coefficients, decay coefficients, mass transfer coefficients, and boundary conditions. These relationships and parameters can be determined from direct measurements in laboratory treatability tests that can accompany modeling efforts or can be found through a literature search.

Multiphase models can potentially account for:

1. Advection
2. Dispersion

3. Sink/source mixing
4. Chemical and equilibrium partitioning

The models can potentially simulate the removal of the contaminants from the subsurface under a variety of conditions (different flow velocity, different well screen positions, different moisture levels, different extraction and passive well locations, different concentration profiles, etc.). The models can allow the user to optimally choose:

1. Well screening positions
2. Well locations
3. Well positioning
4. Vapor flow rate
5. Applied vacuum
6. Soil moisture content

PILOT STUDIES

Pilot studies generally are conducted in order to gather relevant information to design a full scale VES. Field pilot studies gather information regarding the pneumatic flow characteristics of the vadose zones and the extracted air quality. Laboratory soil column and soil cube tests allow for simulation of the vapor extraction remediation process by passing air through a small amount of soil. This also allows for selection of the minimally acceptable air flow velocity for removal of the contaminants.

LABORATORY STUDIES

Laboratory studies often are the best method to optimize the required air flow and moisture content, as well as other control parameters, for a VES. Laboratory studies allow for manipulation of these parameters under controlled conditions and can be done in conjunction with modeling and field pilot studies to optimize system performance. Due to the expense associated with these activities, their implementation is rare except at large sites contaminated with nonbiodegradable VOCs where volatilization must be optimized.

Lab studies can be conducted using columns or soil cubes. Figure 13 presents a schematic of the two configurations. Soil cubes offer the benefit of providing better simulation of actual air flow profiles in the subsurface.

Researchers have increasingly concentrated on evaluating the minimally acceptable air flow velocity for successful vapor extraction (Armstrong et al., 1994; Gierke et al., 1989; Sleep and Syikey, 1989; Brusseau et al., 1991). It is anticipated that upon completion of sufficient lab studies, a matrix of contaminant type, soil type, moisture content, and minimal air flow velocity can be compiled. This matrix would greatly improve the design of VES by practitioners.

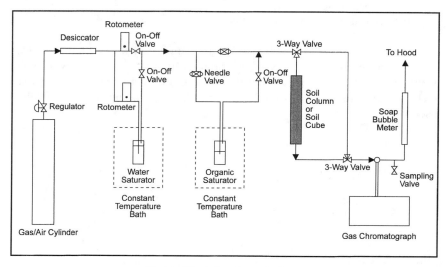

Figure 13 Lab columns/cubes.

FIELD PILOT STUDIES

This section will describe the test wells, procedures, and equipment required to conduct a pilot or feasibility test. A pilot test will allow determination of the subsurface air permeability and the expected mass removal rates of contaminants at system startup and will allow for the selection of appropriate vapor treatment equipment. The pilot test also should determine the required number of extraction points to achieve the desired air flows during VES application. A pilot test should be conducted prior to installation of most VES applications. Only systems whose size does not greatly exceed (2 to 3 times) the size of the pilot system should be installed without pilot testing. For small systems, the cost of the pilot test sometimes cannot be justified.

A typical pilot test setup is shown in Figure 14. The pilot test will include installation of a minimum of one extraction well, several observation wells (or points), and hookup of the extraction well to the vacuum equipment. Upon startup of the vacuum pump, several field measurements from the wells and the extraction vacuum pump are taken prior to the conclusion of the test.

Vapor Extraction Testing Well

The extraction well should be located near the center of the contaminated zone in order to ensure gathering of data that may be representative of startup conditions. The vapor extraction well should be 2 to 4 in. in diameter. At most locations, 2-in. diameter construction is adequate; 4-in. diameter construction is advisable in instances where groundwater pumping is considered from the vapor

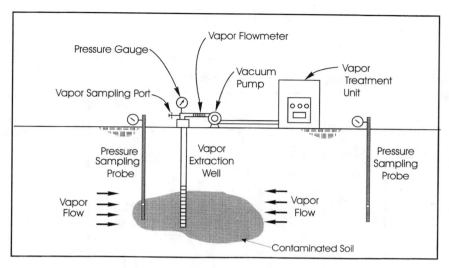

Figure 14   Pilot test setup.

extraction well (vacuum enhanced recovery or dual extraction) or for deeper contaminated zones where significant air flow may be generated.

In some instances, existing monitor wells also can be used for vapor extraction; however, water uplifting should be expected to reduce the available screen length. Should uplifting significantly reduce the available screen length, installation of a proper vadose zone extraction well should be undertaken.

The screened interval of the extraction well should be within the contaminated zone. Since air circulation below the screened interval generally is not significant, the screen should be placed accordingly to attain the desired air distribution in the contaminated zone. Where contamination is deep and permeability is high throughout the soil column, the screened section should be extended to the maximum depth possible and slotted only at the bottom to maximize the area of treatment (rather than slotted fully vertically) (Sims, 1990).

The design of a typical VES vertical well is shown in Figure 15. Horizontal wells may also be considered in shallow water-table situations. Horizontal wells allow for vacuum distribution over a larger screened area to minimize water uplift. Short-circuiting concerns are generally amplified with shallow horizontal wells; therefore, extraction well vaults generally are not installed at the well. Piping is usually run back to a point farther away from the extraction lateral where the vault can be safely installed.

Vapor Extraction Monitoring Wells

A pilot test will generally require three to four monitoring points in a homogeneous setting to assess the zone of air flow influence. The number of required monitoring points increases if the site is nonhomogeneous or nonisotropic. The

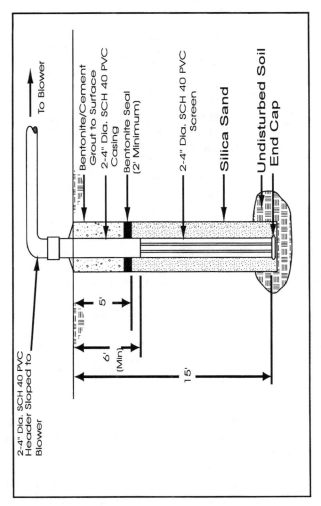

Figure 15 Typical VES well.

increased number of points depends on the complexity of the site. It is best to locate at least one point close to the VES well in order to ensure one positive result point (10 ft). Other points are located at increasing horizontal (and vertical, if required) distances. Often one tries to locate the test in an area where existing monitoring wells can be utilized for observation.

Monitoring points often can be existing groundwater monitoring wells if they are adequately located and screened. Alternatively, monitoring wells of similar construction to the extraction well can be installed. If the observation wells will only be used only for vacuum measurement, they may be of small diameter (5/8 to 1 in.); however, if one desires to gather *in situ* air velocity profiles with a down-well anemometer, appropriately sized wells are required. In shallow conditions, and instances where the soils are sandy and not silty, the observation points may be driven (5/8-in. soil vapor probes can be used). Driving points in silty conditions may lead to clogging of the drive point screen.

The pilot test can be subdivided into three stages: (1) pilot test planning, (2) conducting the pilot test, and (3) data evaluation. These stages are described below.

Stage 1: Pilot Test Planning

In addition to considering the contaminant and site characteristics prior to conducting the VES pilot test, there are several issues which need to be addressed:

1. *Well location and construction:* The extraction well must be adequately screened and located to achieve the desired results. The screen should be in the contaminated zone and groundwater uplift must be given consideration. It is generally advisable to locate the well in the middle of the contaminated zone.
2. *Short circuit pathways:* Short circuit pathways should be minimized. Wells should be located and screened to minimize the short-circuiting effects of high permeability utility trenches, open surfaces, or other low-resistance flow pathways. For example, a well screen which spans two geologic sections will collect most of the air flow from the more permeable section. If the air flow is desired in the low permeability formation, the high permeability zone will act as a short circuit pathway.
3. *Required test duration:* The pilot test must evacuate a minimum of 1.5 to 2.0 pore volumes of contaminated vapors in order to gather vapor quality that would be representative of VES startup conditions. A pore volume can be calculated roughly by assuming a radius of capture (40 to 50 ft typical), assuming a soil porosity, and evaluating the time requirement to evacuate that volume of air. Upon removing the first pore volume, the VOC concentrations will typically drop and will be more indicative of VOC levels to be observed upon system startup. Soil vapor concentration levels of the first pore volume are indicative

of equilibrium conditions and tend to be higher than observed during VES operation.

4. *Vacuum equipment needs:* You should check Figure 5 for expected vacuum needs based upon the soil conditions of the site and select a blower/vacuum pump that will yield the required air flow at the expected subsurface resistance. Most wells are not 100% efficient, and higher vacuum requirements should be anticipated. A 25% decrease due to well inefficiency can often be expected. Inducing air flow at the pilot test is a requirement; therefore, one must utilize a blower/vacuum pump that can induce air flow in the subsurface at the expected resistance. In the case of uncertainty regarding the vacuum requirements, it is best to be conservative and utilize a high vacuum pump. If the high vacuum is not required, the well-head vacuum during the pilot test will be low, and full-scale system design can include a low-vacuum blower.

5. *Type of monitoring points:* Existing groundwater monitoring wells can be utilized if they are screened in the vadose zone; alternatively, new monitoring points can be installed. Soil vapor probes (5/8 in.) can be driven at various depths (15-ft practical limit) but there is a risk of clogging in silty conditions. Checking the vacuum rebound on a probe is a good test to see if it is clogged. The vacuum rebound test is done by applying a vacuum to the probe with a small vacuum pump and seeing how quickly the gauge rebounds after the vacuum is turned off. A clogged screen rebounds slowly, whereas an unclogged screen bounces back instantly. Probes offer the flexibility of varying location during the test. The use of probes precludes the use of a down-well anemometer to record *in situ* air flow velocity.

6. *Off-gas treatment requirements:* Since the pilot test is of short duration, the choice is generally between granular activated carbon (GAC) or rental of other equipment (regenerable vessels or catalytic/thermal oxidizers). Granular activated carbon consumption (and therefore costs) needs to be evaluated vs. other equipment rental costs to make the selection.

Stage 2: Conducting the Pilot Test

The VES pilot test will include the following measurements or activities:

1. Measure the flow rate of extracted air and well-head vacuum. Vacuum/flow rates can be adjusted by opening and closing a dilution valve on the influent side of the blower/vacuum pump. Opening the valve reduces the subsurface air flow/vacuum application. This variation and the monitoring of the subsurface response is useful for computer model verification and calibration, as well as providing multiple data points for field parameter (permeability) evaluation.

2. Measure vacuum influence in the soil probes (or monitoring wells) that is induced by the extraction well using magnehelic gauges. Down-well

anemometers should be used to monitor well air flow velocities. Anemometers have a limited flow velocity range (should be used closer to the extraction wells) and are very sensitive to moisture buildup. Readings should be taken as quickly as possible to minimize the impact of moisture buildup. The measurement of *in situ* air velocities is not widely practiced, and limited information is available on the best methods for measurement. Figure 16 provides a suggested test measurement setup.
3. Measure the vapor emissions before and after on-site treatment using a field instrument and collect lab samples for field verification. This data should be collected at the beginning, middle, and end of the test. This data will be essential for evaluating vapor treatment options.
4. Measure the oxygen and carbon dioxide content of the withdrawn air. As will be discussed in the bioventing section, this data is used to assess subsurface biodegradation rates.

Stage 3: Evaluating the Data

The field data is evaluated to determine or select the following.

1. The pneumatic permeability of the site
2. The location of extraction points in the full-scale remedial design to achieve the required air flow distribution in the subsurface
3. The appropriate extraction blower/vacuum pump
4. The appropriate air treatment technology

The first two analyses are discussed below. Chapter 7 of this book discusses vapor treatment, and vacuum equipment selection is discussed later in this chapter.

Evaluation of Pneumatic Permeability

The Jacob equation is often used to evaluate pilot test data. The pneumatic form of the Jacob equation is

$$Q/H = (k/u)P_w \{1-(P_{atm}/P_w)2\}/\ln (R_w/R_i) \qquad (11)$$

where:

k	=	soil permeability to air flow (cm^2) or Darcy
u	=	viscosity of air = $1.8 \times 10E - 4$ g/cm-sec or 0.018 cp
P$_w$	=	absolute pressure at extraction well atm
P$_{atm}$	=	absolute ambient pressure = $1.01 \times 10E 6$ g/cm-sec^2 or 1 atm
R$_w$	=	radius of vapor extraction well (cm)
R$_i$	=	radius of vacuum influence of the vapor extraction well (cm)

Equation 11 is not very sensitive to values of R$_i$ and, therefore, the subjective nature of the value does not alter the evaluation significantly. Evaluation of the

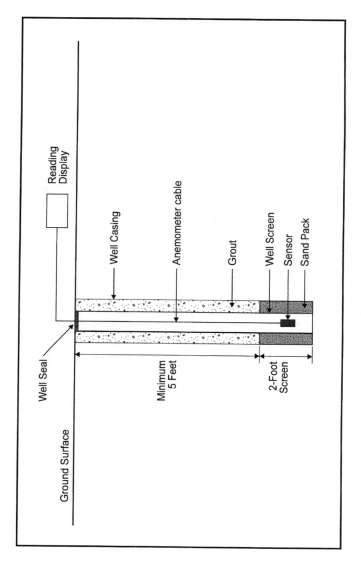

Figure 16 Anemometer setup.

pneumatic permeability using Equation 11 allows for calculation of air-flow velocities across the site as will be discussed in the next section.

Extraction Well Placement

Using the pneumatic permeability (k) value, vacuum profiles can be converted into velocity profiles across the site by substitution of (k) into the Darcy and continuity equations for air flow. The Darcy equation is the same as the equation for groundwater flow except that vacuum pressure is the gradient rather than water elevation. The continuity equation simply states that mass entering a unit volume must also exit the unit volume (if biodegradation does not occur in the unit volume). This task is best accomplished using numerical modeling to provide velocity profiles across the entire site (using air flow models). This translation can also be compared with the *in situ* velocity measurement using a down-well anemometer during the pilot test.

The velocity profile is used to locate wells across the site to achieve the minimally acceptable velocity in the contaminated zone. Numerical modeling is often beneficial in simulating the effects of multiple extraction points. Minimally acceptable air flow velocities can be determined based upon literature reviews for the particular soil type and contaminant or can be based upon laboratory studies that simulate the subsurface flow/contaminant conditions. Laboratory studies are typically not conducted for small applications or applications where the contaminants are biodegradable (petroleum contamination), since it is the general belief that the inducement of air flow will eventually lead to biological breakdown of the contaminants.

Reliance on minimally acceptable vacuum readings (0.05 to 0.5 in H_2O) or a minimally acceptable percentage of applied well-head vacuum to determine the VES radius of influence is generally an unacceptable method of data analysis for large sites or sites contaminated with nonbiodegradable compounds. This practice however is often conducted at small sites such as service stations where well spacing is generally conservative (typically 40 to 60 ft) and is often limited by physical barriers (tank, piping, and building location) and, therefore, cannot be reasonably optimized. Most of these VES installations have a significant reliance on bioventing, where volatilization velocities are not optimized.

SYSTEM DESIGN

Vapor extraction system design can be categorized as being done with the aid of modeling for larger more complicated projects or done with a more empirical design without modeling for smaller type projects. The design considerations for both approaches are similar and are enumerated below. The use of air flow or multiphase transport models allows for a significantly improved ability to locate extraction wells and to select the most appropriate subsurface air flow velocities. Models allow the designer to design a system with greater confidence that it will achieve its remedial goals. Empirical design methods rely on past experience or on achieving air flows based upon predetermined vacuum levels in the subsurface. This approach generally yields lower confidence levels that remedial goals will be achieved. The designer must decide whether modeling costs and the higher confidence levels warrant the added cost.

The following design considerations are enumerated as a practical checklist guide for the designer.

1. *Air-flow rates* (air-flow rates to achieve remedial goals within the required time frame): Air flow requirements will dictate vacuum equipment selection and air emission control equipment sizing. Air flow requirements are dictated by modeling, pilot testing levels, and/or literature review to determine adequate air flow across the entire site.

2. *Vacuum equipment selection:* The required vacuum to induce the desired air-flow rates dictates the type and size of vacuum pumps or blowers that are required. This information is generally based upon a scaling-up of the field pilot test results. In general, regenerative-type blowers typically are used in low-vacuum (to 8 inHg) and high-flow applications; positive displacement blowers are used in medium-vacuum applications (to 12 inHg) and liquid-ring pumps are used to induce flow at high resistance applications (to 25 inHg). Other types of blowers and vacuum pumps such as rotary vane and gear pumps have been used successfully for VES but tend to be less common. Due to the continuous operation of VES, vacuum equipment should be protected by thermal overload shut-offs. Vacuum equipment systems always should be designed with dilution valves in order to adjust the applied vacuum and to dilute the recovered vapors for vapor treatment, if required.

3. *Knockout tanks/fillers:* Blowers/vacuum pumps should be protected from suction of water and particulates by using in-line filters and moisture knockouts. In instances where significant moisture is accumulated (high vacuum or dual extraction), the accumulated water in the knockout is often transferred to the groundwater treatment system using a transfer pump.

4. *Measurement devices:* VES should be equipped with both flow measurement and vacuum measurement devices. Vacuum gauges should be placed on the suction side of the blower.

5. *VES well design:* VES wells should be vertically installed as shown in Figure 14. Lateral well installation can be considered for high water table applications. Wells should always be equipped with flow adjustment valves and vacuum gauges. Valves allow for process control by fully or partially utilizing the extraction point during the remedial process. The use of trenches for vapor extraction is generally a poor choice due to the limited ability to control flow pathways and the high risk of short-circuit pathways within trenches. Unlike wells, trenches do not allow for valving down or closing certain areas of the site during the cleanup process.

6. *Number and location of VES wells:* The number of VES wells will be dictated by the need to maintain adequate air flow in the subsurface. The pilot test in conjunction with the computer simulation will determine the number of required wells and their ideal locations.

7. *Passive well placement:* Passive wells provide influent air and should be located to minimize "dead" (no flow) zones that may develop in multiple extraction point configurations. Passive wells can also be utilized to provide engineered short circuit pathways so as to eliminate migration of contaminants from certain zones (possibly off-site). See Figure 17.

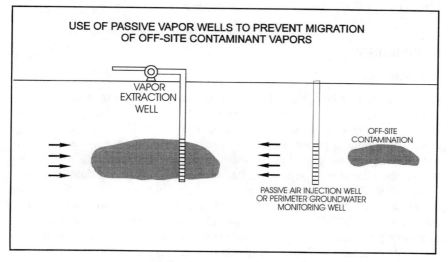

Figure 17   Passive wells/cutoff wells.

8. *Well screen positioning:* Screened intervals are particularly important in ensuring that advective flow is induced in the most contaminated zones. During vapor extraction from thick vadose zones, the screen should be placed near the bottom of the vadose zone rather than through the entire zone.

9. *Soil moisture content:* Soil moisture controls air permeability and may control contaminant partitioning and VES rate limitation. Its proper manipulation may enhance system performance. If one cannot induce sufficient air flow for volatilization, one may want to consider keeping the moisture content high and enhancing bioventing for biodegradable compounds.

10. *Off-gas treatment technology selection:* Pilot test data, cost analysis, and operation and maintenance logistics will enable the appropriate selection of vapor treatment technologies. See Chapter 7 for more details.

11. *Piping:* All underground piping should be vacuum- or pressure-tested prior to burial. Schedule 40 PVC is generally an acceptable material of construction for most underground piping systems.

Figure 18 shows a schematic of a typical VES design. In many respects, the design of a VES is quite complex. The designer must achieve the optimal air flow velocities in a complex subsurface that is likely to be nonhomogeneous. In many other respects, the above-ground components of a VES are so simple that the designer can forget about the many subsurface uncertainties. The above-listed design considerations simply provide the issues to be considered but, unfortunately, cannot shed light into the nonhomogeneous subsurface.

BIOVENTING

Introduction

Bioventing, or bioenhanced soil venting, is a remedial method similar to soil venting in that it relies upon an increase in the flow of air through the vadose zone by pumping soil air from a well completed in the vadose and capillary fringe zone. Soil venting is performed to volatilize the hydrocarbon constituents *in situ*. In bioventing, the increase in the flow of air is to provide oxygen in the subsurface to optimize natural aerobic biodegradation, and this becomes the dominant remedial process. Bioventing has the advantages of being one of the few *in situ* technologies for remediating both volatile and nonvolatile fractions (as long as the compounds are biodegradable). In addition, bioventing can minimize air treatment prior to re-injection (if used) and has been shown to reduce the contaminant concentrations more rapidly than other methods.

The bioventing technology optimizes the air-flow rate to minimize volatilization while providing sufficient air flow to enhance biodegradation. This will often eliminate or minimize air treatment requirements. During implementation of bioventing, soil gases at monitoring locations (O_2, CO_2, CH_4, etc. — not just system vent gas composition) generally are monitored to ensure the presence of aerobic conditions. Moisture and nutrient addition can be considered to enhance system performance. Both moisture and nutrients must usually be delivered in the aqueous state. Providing coverage of the vadose zone with solutions is difficult and requires a sophisticated subsurface distribution system. Ammonia delivery theoretically can be delivered in the gaseous state, although this has not been commonly practiced. Some recent work has also looked at providing phosphorous in a gaseous form.

Advantages of Vapor Phase Biotreatment

The bioventing system has significant advantages in comparison with groundwater-based aerobic biodegradation systems. By using air as the oxygen source, the minimum ratio of air pumped to the contaminants is approximately 13 lb of air per pound of contaminant for typical petroleum contamination. This compares

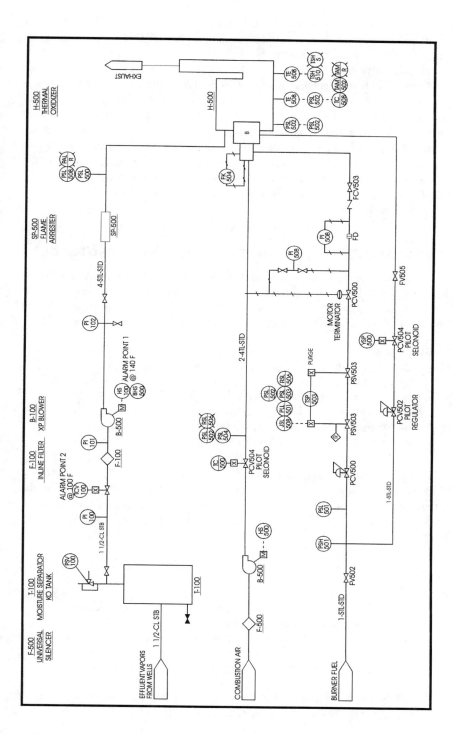

Figure 18 Typical VES above-ground system.

to a requirement of delivering over 1000 gal of ground water to deliver the same amount of oxygen to the contamination. This is because oxygen saturation in water is roughly 8 mg/L, whereas air is 20% oxygen. Another major advantage is that gases have much greater diffusivities than liquids.

Geological heterogeneities present a particular problem for water-borne oxygen flow, since the ground water will be channeled to the more permeable pathways or channels. As a result, oxygen delivery to the less permeable zone must occur by diffusion. If air is the oxygen carrier, the diffusion can take place several of orders of magnitude faster than in the liquid phase. By dewatering and exposing additional vadose zone, the bioventing process can be conducted in the once saturated zone, significantly expediating the remedial process. Chapter 5 will discuss the benefits of transforming the contaminant carrier from water to vapor in greater detail.

PERFORMANCE CRITERIA AND BIOVENTING PLAN PROTOCOLS

Bioremediation projects often have been evaluated by determining whether the contaminant levels are decreasing and whether the microorganisms from the site have the capability to metabolize the contaminant when removed to the laboratory. This simplistic evaluation is inadequate for the following reasons.

- Field microbial activity may not behave in the same manner as the laboratory cultures. In the laboratory, the field microorganisms are put into contact with the contaminant of interest under laboratory conditions that may not simulate natural conditions.
- Biotransformation or other abiotic processes (such as volatilization) may cause reduced contaminant concentrations without actually resulting in biological breakdown of the contaminant mass.

In order to demonstrate that bioventing or biological activity is occurring at a site, sound scientific logic must be demonstrated. This is particularly true due to the difficult credibility route *in situ* bioremediation has suffered in the industry. The following three performance objectives must be met (Rittman and McDonald, 1993):

1. Documented loss of contaminants from the site by sampling and chemical analysis.
2. Testing that shows microorganisms in the laboratory assays have the potential to transform the contaminants under the expected site conditions.
3. One or more pieces of evidence showing that the biodegradation potential is realized in the field: the simplest test to conduct in the field is to measure the electron acceptor uptake rate (oxygen, under aerobic conditions) and to measure the subsurface inorganic carbon production rate (respirometry test).

The laboratory and field biotreatability testing are integral components of demonstrating that bioventing is a viable technology at a given site. The following sections outline the procedures to be followed for conducting the various demonstration tests. The field respirometry testing satisfies the third performance objective. Laboratory treatability testing for the contaminants of interest satisfies the first objective.

LABORATORY TESTING

A series of analyses is performed on site soil samples to evaluate the potential for biological degradation of the contaminants. The evaluations are conducted to determine if the respective soil samples harbor microbial populations capable of using the components of contaminants as carbon sources with the possibility of enhancing the populations to remediate the source material. Both impacted and nonimpacted samples from the site are often used in the evaluation.

The objectives of the lab studies are to (1) determine if aerobic microbes are present in the samples, (2) determine if the microbes have adapted to degrade the selected organic compounds, (3) determine if the environmental conditions (pH and moisture content) are conducive to support microbes, and (4) determine if soluble inorganic nutrients (such as ammonia and phosphate) are present in sufficient quantities for bioremediation of the contaminants.

Total heterotrophic aerobic microbes generally will be performed using the spread-plate procedure. Heterotrophic aerobic microbes capable of degrading specific contaminants can be determined using modifications or variations of the above procedure. Soil respiration testing subsequently can be determined using a respirometer. The concentrations of oxygen and carbon dioxide in the soil chamber headspace are measured periodically during a 1-day period. The difference in oxygen and carbon dioxide concentrations is subsequently graphed and may be correlated with the rate of respiration.

The soil pH, soil moisture, soluble ammonia, and ortho-phosphate concentrations are also generally determined according to accepted standard methods. The contaminant concentrations in the soils are also analyzed according to the relevant and accepted analytical methods in order to quantify the observed degradation in terms of contaminated mass.

If the contaminants in the soil are not known to be biodegradable or system parameter (soil moisture/nutrients, etc.) manipulation is considered, a more sophisticated treatability test can be conducted using soil columns or microcosms. Lab studies involve utilizing multiple microcosms of soil samples in order to vary the parameters of interest (moisture, air flow, nutrients, etc.). The soil within the microcosms is analyzed at various points in time to assess degradation of the contaminants of interest. Parameters such as moisture, pH, and nutrient levels are also monitored during the test period. Analytical costs for these tests can be in the range of $10,000 to $50,000. These treatability tests, therefore, are not conducted for compounds known to be biodegradable, nor at sites with reasonably good nutrient and moisture conditions.

FIELD RESPIROMETRY TESTING

The respirometry test consists of ventilation (introduction of oxygen) of the contaminated area and periodic monitoring of the depletion of oxygen and the production of carbon dioxide for a periodic of time (3 to 5 days) after the air source is turned off. Based upon the results of the respirometry test, oxygen uptake rates and biodegradation rates can be approximated. The oxygen uptake rate subsequently can be utilized to optimize air flows in the subsurface.

The typical test setup is shown in Figure 19. The respirometry test procedures have been best documented by Hinchee and Ong for several demonstrations at U.S. Air Force bases (Hinchee and Ong, 1992a; Hinchee and Ong, 1992b). The monitoring points typically are narrowly screened in the zone of interest (contaminated area). This is because oxygen concentrations in the typical monitoring well may not be representative of local conditions. The narrow screening ensures measurement at precise locations. Air is typically injected in one to five points at flow rates in the range of 1 cfm for a period of 24 h for a typical 10 to 20 ft vadose zone. Subsequently monitoring point gas composition is analyzed for 2 to 4 days. The monitoring points can be as simple as soil gas monitoring probes (5/8-in. diameter steel with a screened 6-in. section) or slightly larger diameter 2-in. monitoring wells. It is generally advisable to mix 1 to 2% helium with the injected air as a tracer gas. Helium can be easily monitored and detected with field instruments (with accuracy to 0.01%). Detection of helium implies that the gas sampled is the same gas that was injected and that changes in its makeup (O_2/CO_2/contaminant distribution) are attributable to bioactivity. A relatively constant detection of helium concentrations over the monitoring period implies that the gas injected is the gas sampled.

After injection of the air/helium mixture, the soil gas is periodically monitored at a frequency consistent with the oxygen uptake rate (every 2 to 12 h). The test is normally terminated when the oxygen concentration drops to 5% or after 3 to 5 days. Oxygen utilization rates can be determined from the slope of the O_2% vs. time graph, if a zero-order respiration rate is assumed. Zero-order rates have been observed for jet fuel degradation (Hinchee and Ong, 1992a; Hinchee and Ong, 1992). As a first approximation to estimate the biodegradation rate, the stoichiometric relationship for contaminant consumption can be formulated. The following example illustrates the calculations for consumption of benzene:

$$C_6H_6 + 7.5\ O_2 \longrightarrow 6\ CO_2 + 3\ H_2O \tag{13}$$

Using the oxygen consumption rate calculated from the oxygen concentration/time graph, the biodegradation rate in terms of milligrams of benzene-equivalent per kilogram of soil per day can be estimated using the following relationship:

$$K_D = -K_R\ V\ D_o\ C/100 \tag{14}$$

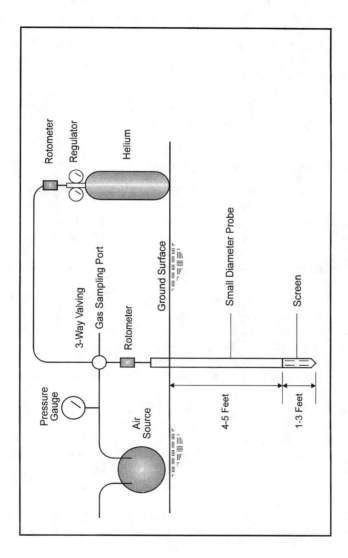

Figure 19 Biovent test setup.

where:

K_D =   biodegradation rate (mg/kg/day)
K_R =   respiration/oxygen utilization rate
V =   volume of air per kilogram soil (L/kg)
D_o =   density of oxygen (1330 mg/L; temperature, pressure specific)
C =   mass ratio of benzene to oxygen required for degradation (78/240 = 0.32)

Based upon a porosity of 0.3 and a soil density of 1440 kg/m³, K_D can be estimated to be roughly 0.89 times the measured respiration rate (K_R) for benzene. Alternate methods that evaluate biodegradation rates based upon CO_2 generation may be less accurate than oxygen consumption-based calculations, if carbonate precipitates may be formed from the gaseous CO_2 production. Formation of these precipitates is dependent on subsurface pH and alkalinity.

SOIL GAS PERMEABILITY TESTING

The pneumatic permeability testing procedure for bioventing is very similar to the pilot testing procedures for VES design. The analysis for zone of influence is based upon providing sufficient oxygen delivery to the location most distant from the extraction/injection point rather than providing adequate air flow for volatilization. Air-flow requirements generally are much lower for bioventing than for vapor extraction; therefore, the notion of using vacuum/pressure zone of influence for well spacing becomes more acceptable and less prone to error than for VES design. If a vacuum is observed, then some air flow is being delivered to that location. Despite this simplistic assessment, a calculation to evaluate whether this vacuum/pressure delivers the required oxygen should be conducted to ensure optimal system performance. This will require translation of the vacuum profiles across the site into air flow velocities. The air flow velocity in turn can lead to an oxygen delivery calculation based upon 20% oxygen content in air. This delivery rate can be compared to the respiration rate. Since many areas in the site will receive oxygen-depleted air, the oxygen delivery calculation yields best-case oxygen delivery information. A pore volume calculation for the zone of influence (based on vacuum influence), and the subsequent translation of the oxygen delivery through the zone of influence provides an alternate method of calculating whether sufficient oxygen is being provided. For example, passage of 1 pore volume per week can lead to evaluation of the oxygen content of this pore volume. Subsequently, one can estimate the required number of pore volumes of air to deliver sufficient oxygen to degrade the adsorbed contamination. Since only some of the delivered oxygen is consumed, more than the stoichiometric oxygen needs will require delivery.

BIOVENTING SYSTEM CONFIGURATIONS

Bioventing systems can be operated to extract, inject, or extract and inject air. Some researchers have noted that high vacuums (extraction mode) affect microbial growth; therefore, some practitioners prefer the air-injection mode of operation. Since high vacuums (5 inH$_2$O) are observed near the extraction well, this is not likely to be a significant issue. Alternating system well operation from injection/extraction mode is also often used to overcome this concern. Figure 20 illustrates several system configurations.

CLEANUP GOALS AND COSTS

Although vapor extraction allows for significant mass removal of VOCs from the subsurface and can be orders of magnitude more efficient than pump and treat remediation, due to the higher transport abilities of the air carrier, reasonable cleanup goals still must be established. Vapor extraction systems reach asymptotic cleanup levels due to nonequilibrium partitioning such as desorption, pore diffusion, or other rate-limiting transport steps that eventually render the system diffusion process limited (Figure 4). In order to reach the cleanup criteria, the end points must be realistically determined. Due to its ability to rapidly remove large amounts of VOCs, vapor extraction is an excellent source control remedial strategy. As has been previously discussed, VES efficiency declines as the remaining contamination mass declines. This tailing effect implies that alternate strategies should be considered during the end of the life cycle of the project. The alternate strategies may include a reduction in the extracted air volume and a greater reliance of passive venting or bioventing. Alternatively, upon completion of the achievable source control (assymptotic levels), natural attenuation of the receptor groundwater may become an acceptable remedy.

Realistic expectations of system achievements are essential in order to minimize operation and maintenance costs. Table 2 provides typical costs to be expected with installation of a VES system. Systems should be operated to the point where the total mass minimizes health and environmental impacts.

CASE STUDY

A 1/2-acre site in Vermont was affected by a 3000-gal gasoline spill. The site was overlain by 10 ft of sandy fill material atop a highly fractured shale. The 10-ft thick fractured shale formation was atop a more competent bedrock. Both ground water and soil within the fractured shale and the overburden were heavily contaminated by the gasoline release. A small sheen of NAPL was present in most on-site wells. The groundwater table is encountered at an approximate depth of 8 ft. The competent bedrock appeared to be minimally contaminated by any dissolved VOCs. The site remedy called for dewatering the fill material and the fractured shale with subsequent vapor extraction of the dewatered soils. Pilot testing was conducted at the site and indicated that the subsurface could sustain an airflow of 200 cfm from eight vapor extraction wells at a resistance of 5 inHg. The pilot test was conducted after 50% site dewatering had been accomplished by pumping from two on-site groundwater recovery wells for 3 days. The

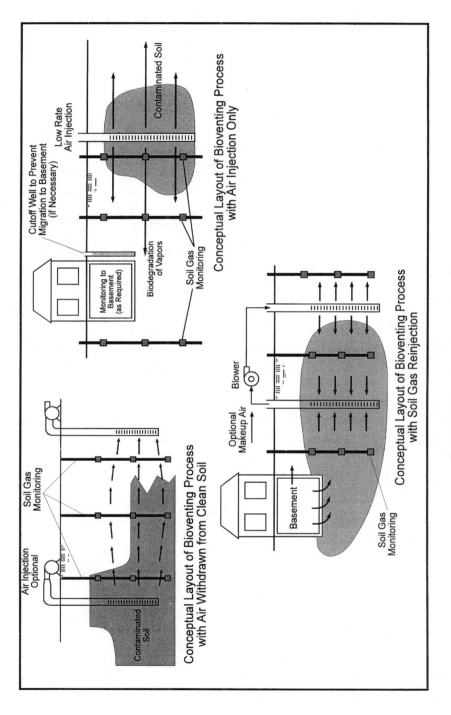

Figure 20 Bioventing system configurations.

Table 2 Typical System Component Costs for Vapor Extraction

	Item	Cost ($)
1. Blowers/vacuum pumps (100 cfm)		
1A	Regenerative type	3000
1B	Positive displacement type	5000
1C	Liquid ring type	6500
2. Filters and knockout systems (100 cfm)		
2A	For regenerative positive displacement type blowers	2000
2B	For liquid ring type pumps	5000
2C	Liquid transfer pump	1000
3. Extraction wells		
3A	20-ft vertical 2-in. PVC well (assume one of three installed per day)	1000
3B	Shallow (6-ft) horizontal extraction well (4-in. PVC) (assume one of three installed per day)	1000
4. Trenching and underground work		
4A	Costs per linear foot (4-in. PVC header), oversight costs included, soil disposal not	20
4B	Concrete manhole per extraction well (4 ft × 4 ft × 4ft)	400
		5000
5. Other internal above-ground components (100 cfm)		3000
5A	Piping/valve (PVC)	2000
5B	Gauges and flow meters	1000
5C	10 ft by 12 ft heated and insulated shed	4000
5D	Controls and electrical	5000
		2000
6. Air Emission control		
6A	100 cfm disposable GAC	2500
6B	100 cfm catalytic oxidizer	30,000
6C	100 cfm thermal oxidizer	30,000
6D	500 cfm regenerable GAC system	45,000
6E	500 cfm regenerable resin system	80,000

extracted vapors during the pilot test (from one well) revealed total VOC levels in the range of 5000 ppm.

The VES system was designed to operate at 200 cfm and to utilize a positive displacement (PD) blower. The airflow was selected based upon cost-benefit analysis. A 200-cfm air stream could be handled with reasonably priced off-gas treatment equipment. Due to the biodegradability of the VOCs, the airflow was not maximized for volatilization, but rather selected to deliver adequate airflow for enhanced biodegradation and remain low enough to minimize off-gas treatment costs. Upon system startup with the eight-well configuration, vacuum was observed at all observation points, indicating the presence of airflow and thereby oxygen delivery for bioventing.

For the above-ground system, prior to the PD blower, a knockout tank and air filter were installed. The knockout tank collected 200 gal of water per hour, and a transfer pump was utilized to pump the moisture or ground water to the groundwater treatment system. Extracted soil vapors were treated using a catalytic oxidizer. Within 24 months, 10,000 lb of hydrocarbons were volatilized from the soils. An additional 10,000 lb were biodegraded based upon monitoring of O_2/CO_2 levels in the off gases. The oxidizer only operated for 6 months, after which time emissions were treated with activated GAC based upon cost-benefit analysis. Extracted vapor levels dropped to 10 ppm within one year, and after 2 years of operation, the levels were roughly 5 ppm, demonstrating the long tailing effects that are seen with many remedial systems. Initial levels of extracted groundwater (10 gpm) showed levels of dissolved BTEX levels in excess of 50 ppm. After one year, the BTEX levels dropped to 1.5 ppm, and the levels after 2 years averaged 900 ppb of dissolved BTEX.

This case history demonstrates the rapid success of vapor extraction and the diminishing mass removal effects of continued operation. The slower pace of groundwater concentration decline demonstrates that groundwater flushing is slower in reaching this assymptotic level as well being an order of magnitude less efficient in total mass removal. Chapter 5 provides greater detail on comparison of mass removal by airflow vs. groundwater carriers.

REFERENCES

Armstrong, J.E., Frind, E.O., and McClennan, R.D. "Non-equilibrium Mass Transfer Between the Vapor, Aqueous, and Solid Phases in Unsaturated Soils During Vapor Extraction." *Water Resourc. Res.,* vol. 30, p. 355, 1994.

Brusseau, M.L., Jessup, R.I., and Rao, P.S.C., "Transport of Organic Chemicals by Gas Advection in Structured or Heterogeneous Porous Media. Development of a Model and Application to Column Experiments." *Water Resourc. Res.,* vol. 27, p. 3189, 1991.

Buscheck, T.E. and Peargin, R.G. "Summary of a Nation-Wide Vapor Extraction System Performance Study," Chevron Research and Technology Company, Richmond, CA, 1992.

Farrell, J. and Reinhard, M. "Desorption of Halogenated Organics From Model Solids, Sediments, and Soil Under Unsaturated Conditions. 1. Isotherms." *Environ. Sci. Technol.,* vol. 28, p. 53, 1994.

Farrell, J. and Reinhard, M. "Desorption of Halogenated Organics From Model Solids, Sediments, and Soil Under Unsaturated Conditions. 2. Kinetics." *Environ. Sci. Technol.,* vol. 28, p. 63, 1994.

Gierke, J.S., Hutzler, N.J., and McKenzie, D.B. "Vapor Transport in Columns of Unsaturated Soil and Implications for Vapor Extraction." *Water Resourc. Res.,* vol. 25, no. 1, p. 81, 1989.

Grathwohl, P. and Reinhard, M. "Desorption of Trichloroethylene in Aquifer Material: Rate Limitation at the Grain Scale." *Environ. Sci. Technol.,* vol. 27, p. 2360, 1993.

Hansen, M.A., Flavin, M.D., and Fam, S.A. "Vapor Extraction/Vacuum-Enhanced Ground-water Recovery: A High Vacuum Approach," Proc. Purdue Industrial Waste Conference, Lafayette, IN, 1994.

Hinchee, R.E. and Ong, S.K. "Test Plan and Technical Protocol for a Field Treatability Test for Bioventing," Document prepared for U.S. Air Force Center for Environmental Excellence, Brooks Air Force Base, TX, 1992a.

Hinchee, R.E. and Ong, S.K. "A Rapid In Situ Respiration Test for Measuring Aerobic Biodegradation Rates of Hydrocarbons in Soil," *J. Air Waste Manage. Assoc.,* vol. 42, p. 1305, 1992b.

Hoag, G.E., Bruell, C.J., and Marley, M.C. "Induced Soil Venting For Recovery/Restoration of Gasoline Hydrocarbons in the Vadose Zone," *Oil in Freshwater,* 176 (1989).

Johnson, P.C., Kemblowski, M.W., and Colthart, J.D. "Quantitative Analysis for the Cleanup of Hydrocarbon-Contaminated Soils by *In situ* Soil Venting," *Groundwater Monitor. Rev.,* p. 413, 1990.

Johnson, P.C., Kemblowski, M.W., Colthart, J.D., and Byers, D.L. "A Practical Approach to the Design Operation and Monitoring of *In situ* Soil Venting Systems." *Groundwater Monitor. Rev.,* 1991.

Jury, W.A., Russo, D., Streile, G., and Abd, H.E. "Evaluation of Volatilization of Organic Chemicals Residing Below the Surface," *Water Resourc. Res.,* vol. 26, p. 13, 1990.

Lyman, W.J., Reehl, W.F., and Rosenblatt, D.H. *Handbook of Chemical Property Estimation Methods,* New York: McGraw-Hill, 1982.

Mahan, B.H. *College Chemistry,* Boston: Addison Wesley, 1966.

Piwoni, M.D. and Bannerjee, P. "Sorption of Volatile Organic Solvents from Aqueous Solution onto Subsurface Solids," *J. Contam. Hydrogeology,* vol. 4, p. 163–179, 1989.

Rittman, B. and McDonald, J.A. "Performance Standards for Bioremediation," *Environ. Sci. Technol.,* vol. 10, p. 27, 1993.

Sims, R.C. "Soil Remediation Techniques at Uncontrolled Hazardous Waste Sites, A Critical Review." *J. Air Waste Manage. Assoc.,* vol. 40, p. 704, 1990.

Sleep, B.E. and Syikey, J.F. "Modeling of Transport of Volatile Organics in Variable Saturated Media." *Water Resourc. Res.,* vol. 25, no. 1, p. 81, 1989.

Thibaud, C., Erkey, C., and Akgerman, A. "Investigation of the Effect of Moisture on the Sorption and Desorption of Chlorobenzene and Toluene From Soil." *Environ. Sci. Technol.,* vol. 27, p. 2373, 1993.

5 VACUUM-ENHANCED RECOVERY

Peter L. Palmer

INTRODUCTION

Although vacuum-enhanced recovery has been used for decades as a standard approach for dewatering and stabilization of low permeability sediments or to speed dewatering of more permeable sediments, it has not been until recently that it has been incorporated into groundwater remediation applications. The use of vacuum-enhanced recovery systems in environmental remediations is unique because whereas most remediation methods rely on either water or air as the carrier, vacuum-enhanced recovery relies on a combination of both as carriers. Secondly, vacuum-enhanced recovery uses a combination of two forces, gravity and pressure differential, to move the water. This can be very beneficial in enhancing cleanups when used in the proper hydrogeologic setting.

In Chapter 1 we discussed how gravity is used as a main force to move water in all methods that use water as a carrier. The main method to control the direction of water movement is to use a well to remove water from the aquifer. This creates a drawdown of the water in the aquifer; water travels from a place of high head (high water level) to a place of low head (low water level at the bottom of the pumping well). The liquid state of water allows this vertical force to create a horizontal movement of the water. This principal has been successfully practiced as a remedial technique in the medium and high permeability geologic formations. The success of this technique (using gravity alone as the main force), however, can be severely restricted in lower permeability formations due to the diminished groundwater flows that can be achieved by standard remedial recovery equipment. Vacuum-enhanced recovery systems overcome this limitation by using a second force, pressure differential, to help the movement of the water when gravity movement is limited by the geology. Vacuum is applied, in addition to pumping, to move the water. This combination allows us to move air and water in geologic formations that were inaccessible before.

0-87371-995-6/96/$0.00+$.50
© 1996 by CRC Press, Inc.

Within this chapter we will demonstrate the value of applying a vacuum to improve the performance of a well in moving water in lower permeability geologic settings. However, vacuum-enhanced recovery systems are also advantageous in lower permeability formations using air as a carrier of many organic contaminants. Lower permeability formations generally have silts and clays incorporated into the matrix and these are generally very adsorptive of organic contaminants, and use of water alone as the carrier makes it difficult to achieve removal of these contaminants. However, air is a very effective carrier of many of these organic constituents, and vapor-enhanced recovery provides a mechanism to incorporate air as a carrier in these lower permeability formations.

To overcome the air and groundwater flow restrictions of low permeability formations, high vacuums are created at a well by liquid-ring pumps or other specialty pumps. When coupled with recovery wells, these are collectively referred to as vacuum-enhanced recovery systems. These specialty pumps are used to create high vacuums (up to 24 inHg as opposed to 3 to 6 inHg for conventional vacuum blowers) which results in a much greater driving force (pressure differential) for air flow in the unsaturated zone and, when combined with gravity, increases the rate of groundwater recovery and the size of the groundwater recovery capture zone. In this chapter we will discuss the geologic setting and constituent types that are most suitable for vacuum-enhanced systems and we will take you through the steps needed from evaluation of system applicability to system design.

MASS BALANCE APPROACH TO SITE REMEDIATION

Since vacuum-enhanced recovery uses both water and air as carriers, it is particularly important to understand which phase the constituents are in so that vacuum-enhanced recovery can be used to its fullest benefits. In a general sense, the transfer of the constituents between phases (vapor, dissolved, free phase, and adsorbed) is affected by the relative affinity of the constituents to each phase which can be evaluated using the constituent partitioning coefficients. The most effective remediation will create subsurface conditions that will drive the interphase transfer towards the phase(s) that allow for the most efficient mass removal. Site remediation can therefore be viewed as implementing perturbations to the subsurface that will drive physical, chemical, and biological processes towards the site remediation goals. For instance, at sites contaminated with petroleum-derived volatile organic compounds (VOCs), analysis of the equilibrium relationships is greatly simplified. Gasoline-derived compounds are generally biodegradable, volatile, and relatively water insoluble. Mass transfer and removal of gasoline constituents can be 10,000 times more efficient in the vapor phase than in the soluble water phase. Perturbations that alter the equilibrium in the subsurface in order to drive interphase mass transfer from the free, dissolved, and adsorbed phases to the vapor phase would therefore be viewed as beneficial to site remediation.

One way to shift from dissolved phase remediation (pump and treat) to the more efficient vapor phase remediation is to dewater the contaminated sediments, remove easily accessible light non-aqueous phase liquids (LNAPLs), and then remove the VOCs by vapor extraction. In high permeability formations, site dewatering can create extremely large volumes of water; therefore, this approach is generally not practical due to cost considerations and you would likely rely on water alone as your carrier. However, in lower permeability formations (hydraulic conductivity ranges from 1×10^{-3} to 1×10^{-5} cm/sec) you are not dealing with large volumes of water, and these can be dewatered to expose formerly saturated soils to vapor extraction. Fine-grained sediments associated with lower permeability sediments also possess greater specific retention due to increased capillary pressure, which can trap water in the soil pore space and impede the needed air flow. Application of high vacuum via a vacuum-enhanced recovery system can generally overcome most soil capillary pressures, thereby allowing for the removal of the water, which results in increased airflow and increased constituent removal.

In summary, vacuum-enhanced recovery can be very beneficial in aquifer remediation when properly applied. In the water carrier phase, it can be used to enhance the physical removal of LNAPLs or dissolved constituents or it can be used to dewater sediments and allow the air carrier phase to: (1) remove constituents absorbed onto fine grained sediments, and/or (2) provide oxygen to enhance *in situ* degradation of biodegradable compounds.

GROUNDWATER RECOVERY ENHANCEMENT

Let us put aside our focus on the advantages of air being a carrier and first focus on enhancing the water carrier aspect. To do this we must first understand that the basic premise behind extraction wells is that the withdrawal of ground water from a central point causes a decline in ground water levels (drawdown) in the vicinity of the well which is referred to as the "cone of depression"; the drawdown induces groundwater flow towards the well. The area within which all of the water flows to the extraction point or well is referred to as the "capture zone" (see Figure 1). The point on the downgradient side of the well where water is no longer captured is referred to as the "stagnation point". The capture zone is elliptical in shape as a result of the effect of the existing groundwater gradient. It is important to note that the capture zone and cone of depression are not the same due to the existing gradient. The cone of depression is circular and will continue to expand as pumping time increases until equilibrium conditions are reached. The capture zone is determined by superimposing the cone of depression on the water table (groundwater levels) and, as shown in Figure 2, is located in the immediate vicinity of the recovery well and upgradient from it. The capture zone determines the area that will be contained or remediated by a given extraction system.

Low aquifer transmissivities limit the capture zone. Low transmissivities result from either a thin saturated thickness or low permeability deposits. In low

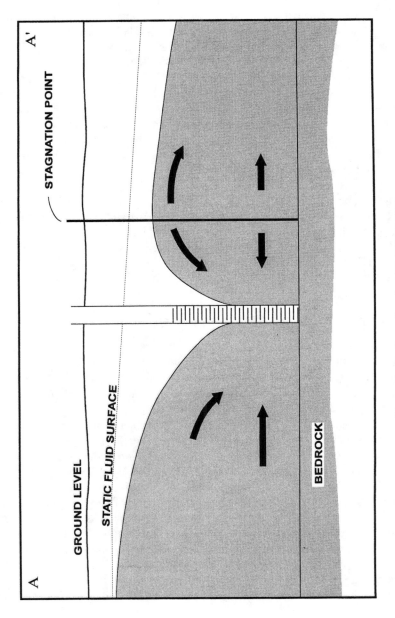

Figure 1 Cross section showing capture zone.

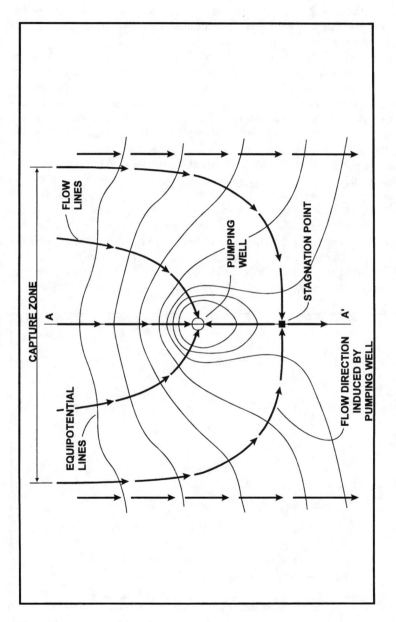

Figure 2 Plan view showing capture zone.

permeability environments, there may be available drawdown; however, the low permeability prevents the formation of a significant capture zone under the influence of gravity alone. In an aquifer with a thin saturated thickness, there is little drawdown available to create a significant capture zone. In these situations, the capture zones are extremely small and the number of wells required using conventional approaches to contain or remediate an area is large.

The capture zone in an aquifer is proportional to the discharge that can be obtained which, in turn, is proportional to the aquifer transmissivity. The discharge for a given transmissivity is limited by the drawdown or gradient that can be produced at the well. Under normal water table conditions, this maximum drawdown is the available saturated thickness and it corresponds to the maximum achievable discharge. If the gradient (drawdown) can be increased beyond the saturated thickness, then the discharge and the capture zone could be increased. This is the basic idea behind vacuum-enhanced recovery — increase the gradient beyond that which can be achieved by pumping alone.

To get this point across, let us first observe the effects of a vacuum on static well conditions. When a vacuum (negative pressure) is applied to the well, the fluid level in the well rises and a "cone of impression" is formed around the well as a result of the negative pressure. This is shown in Figure 3, which shows the water level in the well rising 2 ft above the static water table in response to a vacuum applied at the well. This is a key concept since the negative pressure (represented by a the rise in water levels) increases the available drawdown.

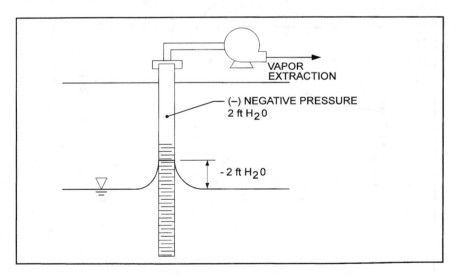

Figure 3   Effect of vapor extraction on water levels.

Now let us look at the effect of a vacuum under pumping conditions. To do this, first look at the effect on water levels from pumping along, as shown in Figure 4, which illustrates a cone of depression formed by an extraction well.

Note that the drawdown at the well created by pumping at 2 gallons per minute (gpm) is 5 ft. When we combine the effects of pumping (Figure 4) with the vacuum effects (Figure 3), the results show that the cone of depression formed by the same extraction well pumping at the same 2 gpm rate is now only 3 ft, due to the applied vacuum (Figure 5). This is what you would actually measure in the well; however, the effective drawdown, which is a combination of the pressure gradient (2 ft) and the liquid gradient (3 ft) is still 5 ft. And, although water level measurements alone would suggest in the second case a smaller capture zone, in fact in both cases the capture zone is the same because the pumping rate remained the same. The gradient required to produce 2 gpm did not change, just the method in which the gradient is created. The benefits of using a vacuum-enhanced system is not just to simply maintain the same pumping rate and capture zone, but to increase the yield of formations (pumping rate) beyond that which could be achieved by pumping alone. This increases your capture zone.

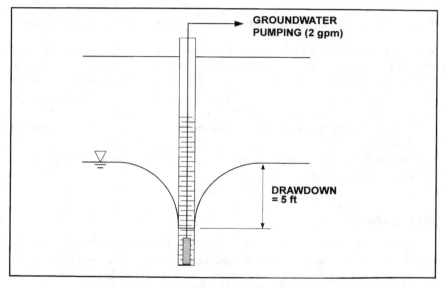

Figure 4   Effect of groundwater pumping on water levels.

Now that we have introduced the concept of effective drawdowns, let us look at what we mean in a little more depth and how it relates to capture zones. The concept of effective drawdown is illustrated graphically in Figure 6, where Q_1 represents the maximum discharge rate that can be obtained by pumping alone due to limited saturated thickness. The figure shows the fluid surface and drawdown associated with Q_1. By applying a vacuum to the well, the discharge rate increases and the capture zone expands as depicted by the drawdown and fluid surface associated with Q_2 in the figure. The combination of the vacuum and the fluid drawdown represents the effective drawdown depicted here. The effective

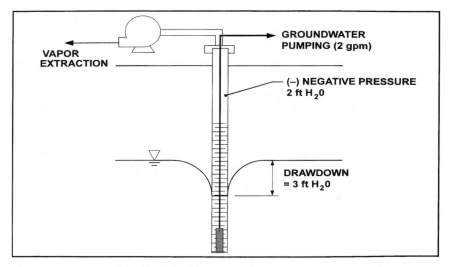

Figure 5 Combined effect of vapor extraction and groundwater pumping on water levels.

drawdown is greater than the saturated thickness due to the vacuum within the vacuum zone (area of measurable vacuum). The increased drawdown outside of the vacuum zone is a result of the increased discharge Q_2. The goal of a vacuum-enhanced recovery system under this scenario is to increase the capture zone; the increased available drawdown resulting from applying a vacuum is the means to achieve this goal. It is easy to see that vacuum-enhanced recovery systems can be advantageous since, when properly used, they can increase the capture zone associated with an individual extraction well, thus decreasing the number of extraction wells needed.

APPLICABILITY

Vacuum-enhanced recovery is applicable to a limited range of hydrogeologic parameters and settings. This is very important, because if your site falls outside of these ranges, then the use of this technology should be critically evaluated and in many cases should not be applied. Generally, low transmissivity is a requirement (<500 gallons per day per foot (gpd/ft)) in order to develop a vacuum of sufficient magnitude to have an effect at reasonable air flow rates. Low transmissivity is indicative of low permeability and small saturated thickness. The system generally needs to possess a permeability in the range of 1×10^{-3} to 1×10^{-5} cm/sec. If the permeability is too high, then the system will not work because the high permeability prevents the formation of a significant cone of impression formed by the vacuum. In other words, the effective drawdown is only marginally greater than the available drawdown with pumping alone. If the permeability is too low, even vacuum-enhanced recovery systems cannot create a significant captive zone. In these cases, increasing the permeability of the formation, as discussed in Chapter 8, would be needed to make these formations suitable for

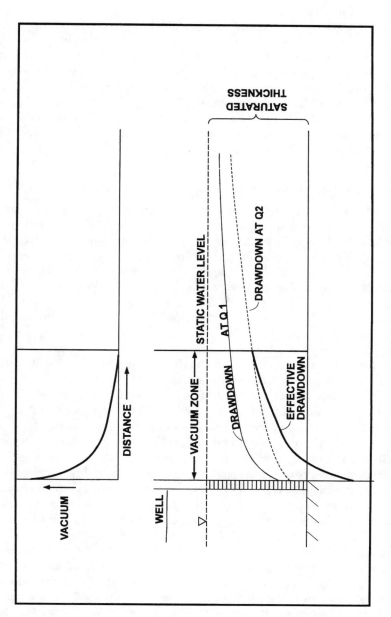

Figure 6 Vacuum effect on water levels.

use for vacuum-enhanced recovery systems. Perched aquifers composed of inter-bedded sands and clays have been demonstrated to be particularly suitable for use of vacuum-enhanced systems. Vacuum-enhanced recovery systems may also be applicable to some fractured systems.

On the management side, there are a number of criteria which could dictate the use of a vacuum-enhanced system. Where suitable, it is generally considered to be a more effective cleanup program with shorter remediation times. In situations where LNAPLs are present, it can avoid source control programs such as excavation and it can be more cost effective from a life-cycle cost perspective than standard technologies due to reduction in capital costs and operating and maintenance costs as a result of shorter remediation periods. Presented below are some applications where vacuum-enhanced recovery was found to be more beneficial than traditional approaches.

Enhanced Effectiveness — LNAPLs

Vacuum-enhanced recovery can be a very effective technology for removing LNAPLs from low permeability, thin aquifers. As we have discussed, conventional pumping methodologies are not very effective because the low permeability combined with little available drawdown limits recovery well capture zones. A good example of where the limited available drawdown was overcome by using a vacuum-enhanced recovery system is at a bulk storage facility (terminal) located on the west coast. At this facility hydrocarbon products were found to be discharging along the bank adjacent to the terminal into a harbor. When the discharges were discovered, daily fines were assessed until the seepage ceased; therefore, the remedial objective was to install a system to prevent further seepage into surface water in the shortest possible time frame. Other remedial options were readily evaluated, such as trenches and barriers; however, they were rejected due to space constraints, costs, soils disposal, and time required for construction. Since the aquifer was thin, with a total depth of 10 ft and a saturated thickness of 8 ft, a vacuum-enhanced system was deemed to be a good possibility.

A preliminary evaluation was undertaken to assess the suitability of using an enhanced vacuum system and the results looked favorable. Consequently, as a next step, a recovery well was installed and a short pilot test was performed to determine the benefits and costs of using a vacuum-enhanced approach compared to a more conventional pumping approach. The pilot study showed that by using a conventional pumping approach, the expected withdrawal rate from a single recovery well would be less than 0.1 gpm, its hydrocarbon recovery rate would be less than 1 barrel per day, and its capture zone in a direction parallel to the bank where seepage was occurring would be small. When a vacuum-enhanced recovery approach was tested on this same well, its withdrawal rate increased to 0.6 gpm, its daily hydrocarbon production rate increased to 27 barrels, and its capture zone increased significantly. The pilot test results demonstrated the benefits of using a vacuum-enhanced approach, and installation of a full-scale system was initiated. Not only did a vacuum-enhanced system allow the use of significantly less recovery wells, Figure 7, the capture zones from a vacuum-enhanced system extended much farther downgradient beyond what could be achieved using conventional pump and treat, so that seepage to the harbor ceased within days of system installation as opposed to weeks using the conventional approach.

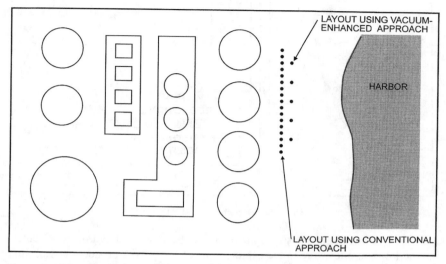

Figure 7 Site layout displaying recovery well layouts using conventional vs. VES approach.

Enhanced Effectiveness — Dissolved Phase

In the above example we focused on a release of LNAPLs from a terminal that required remediation. Vacuum-enhanced systems also can also be beneficial in reducing remediation costs in situations where constituents are in the non-aqueous or the dissolved phase.

Dissolved constituents originating from releases from above-ground storage facilities were discovered at a terminal located in the midwest. The terminal was located adjacent to a river, and a creek separated it from an adjacent industrial facility, which operated on-site production wells (see Figure 8). It was determined that the creek was receiving groundwater discharges from the uppermost sediments, which contained dissolved and liquid hydrocarbons. Discharges to the creek prompted regulatory action, although it was later discovered that the production wells at the adjacent production facility were also impacted.

The cross section shown in Figure 9 shows the two major aquifers, fluid levels in wells, and general flow directions. The upper aquifer consisted of interbedded sands and clays with hydraulic conductivities of the clays ranging from 10^{-4} to 10^{-5} cm/sec. The lower aquifer was extremely productive with a hydraulic conductivity of 10^{-1} cm/sec and was used as a source of water in the area. Several off-site production wells were withdrawing several million gallons of water a day and the pumpage caused a strong vertical gradient in the overlying silty sand/clay aquifer; as a result, pockets of LNAPL within sand lenses were scattered throughout the upper zone.

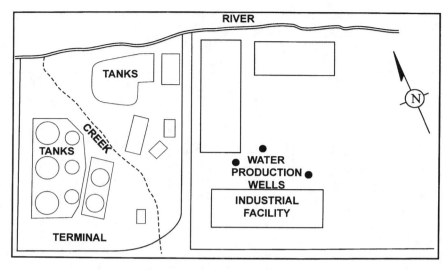

Figure 8 Map showing site layout.

The hydrogeologic investigation confirmed that dissolved hydrocarbons entering the lower aquifer were in the capture zone of the off-site production wells and consequently moved toward these wells. Although the off-site production wells were not used for drinking water purposes, the regulatory agency deemed the risks associated with releases from the site as unacceptable and required that an on-site remedy be developed. One approach would be to install a groundwater recovery system into the lower aquifer; however, this would have required excessive pumping rates (from 1500 to 2000 gpm) and the associated high costs to build and operate a large treatment system. In addition, this approach lacks any source removal, since the LNAPLs would still remain in the upper aquifer and remain as a continuing source of contamination to the lower aquifer.

Consequently, an alternative approach was considered with the goal being to capture hydrocarbon liquids and dissolved hydrocarbons before they migrated into the lower aquifer by installing a recovery system to the base of the upper aquifer. This approach, as illustrated in Figure 10, would be designed to reverse the vertical gradient between the upper and lower aquifers and would cause both the creek and the river to be recharge boundaries (as opposed to discharge boundaries) which would prevent discharges of hydrocarbons into these surface water bodies.

In order to evaluate whether or not a vacuum-enhanced approach would be cost effective, a pilot study was initiated by pumping several recovery wells with and without the aid of a vacuum. The results of the pilot studies showed that under normal pumping conditions, without applying a vacuum, there was limited fluid production and limited influence. However, with the application of a vacuum, the flow rates increased by 5 to 10 times and the water level influence (capture zone) increased from less than 10 ft to up to 100 ft. These studies confirmed the benefits of a vacuum-enhanced recovery approach to intercepting the contamination in the

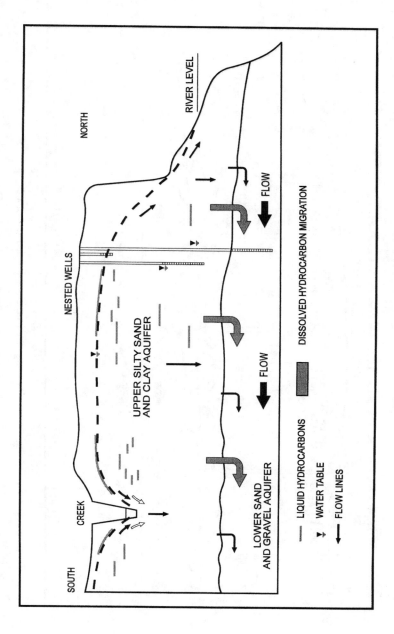

Figure 9 Hydrogeologic cross section.

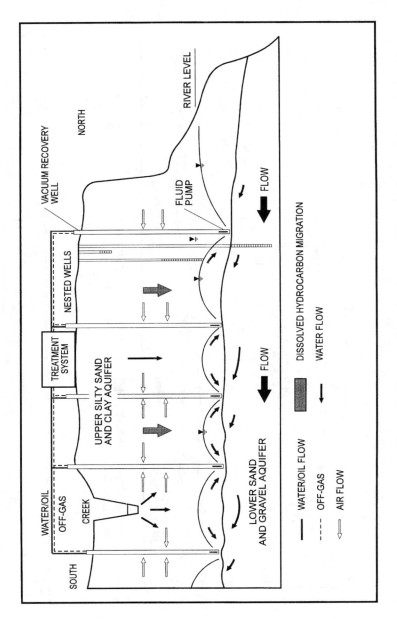

Figure 10 Vacuum-enhanced system remedial approach.

upper aquifer and preventing its movement into the lower aquifer. The pilot studies along with computer modeling showed that a total flow rate of 30 gpm was required to manage the plume and that could be achieved with significantly less wells using the vacuum-enhanced approach. Using air as a carrier to remove VOCs in the unsaturated zone will result in a much shorter cleanup time.

Enhanced Effectiveness — Air Phase

In both of the above examples, the goal of using vacuum-enhanced recovery was to supplement the force of gravity (by adding pressure differentials) to increase the capture zones and recovery rates of individuals recovery wells. In this example, we will show how vacuum-enhanced recovery was used to incorporate air as the main carrier of contaminants. This example is typical of many sites where releases of LNAPLs from underground storage tanks result in the LNAPLs being "trapped" in the sand lenses of interbedded sands and clays, as depicted in Figure 11. The figure shows a release from a site located in the southeast where the LNAPL was confined primarily to a sand seam interbedded within clayey deposits with a permeability on the order of 1×10^4 cm/sec and a transmissivity of 40 gpd/ft. Previous attempts to recover the LNAPL using conventional pumping techniques, i.e., without vacuum enhancement, proved unsuccessful due to the low transmissivity and thin saturated thickness. Consequently, a decision was made to pursue a vacuum-enhanced recovery approach with the objective being to dewater the sand lens and switch to air as the primary carrier to remove residuals LNAPLs that remained in the sand lens after dewatering. Since the LNAPLs were petroleum hydrocarbons, it was anticipated that using air as the carrier would also stimulate bioactivity by increasing the available oxygen to the sand seam.

A pilot test was performed to determine the effectiveness of the vacuum-enhanced recovery approach and to develop design data. The pilot test utilized an explosion-proof liquid ring system capable of moving 50 standard cubic feet per minute (scfm) and producing a vacuum of 24 inHg. Normally, as in the two examples above, you try to maintain high vacuums to maximize flow rates and capture zones. In this example, the objective was to dewater the sand lens and rely on air to achieve the cleanup standards. This is an important difference, since the goal of this remedial program was to maximize air flow and volatilize or biodegrade *in situ* the hydrocarbons by the use of vacuum dewatering. In other words, the objective was to maintain good air flows, not to maintain a high vacuum.

The results of the pilot study showed that the groundwater influence (i.e., capture zone) increased from less than 15 ft using pumping alone to over 100 ft with a vacuum of 12 inHg applied. In addition, the groundwater yield increased from less than 0.1 gpm to 0.15 gpm. More importantly, was the effect on air flow and well-head vacuum as the pilot study progressed which is shown in Figure 12. As the sand lens dewatered, air could more easily flow through it, and the air flow increased with a resulting decrease in well-head vacuum. The air flow increased 50% during the test, and the vacuum decreased 25%. The importance

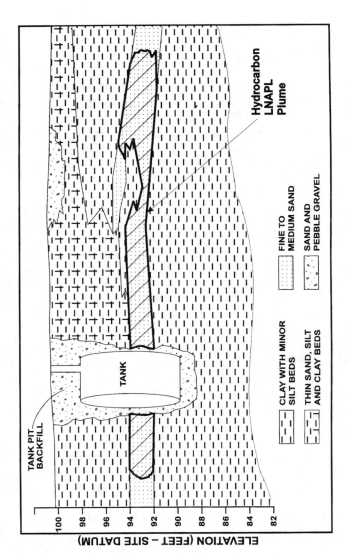

Figure 11   Geologic cross section at release from underground storage tank site.

of maintaining high air flow rates at lower vacuums will become more apparent to you when we discuss equipment selection later in this chapter.

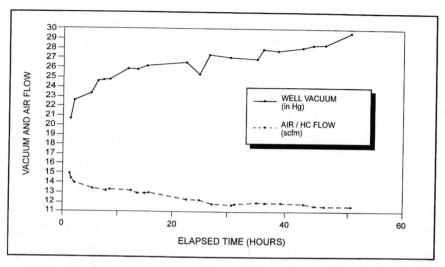

Figure 12   Air flow vs. dewatering during vacuum-enhanced recovery pilot test.

The results of the pilot study also demonstrate the value of collecting concentrations of hydrocarbons in the off-gases, not only to show how effective the system is in volatilizing contaminants, but also to evaluate off-gas treatment requirements. Shown in Figure 13 is the hydrocarbon concentration vs. time in the off-gases; the concentration of about 9% total hydrocarbons declined to about $1^1/_2$% after 50 h of operation. This information is needed during system design (the need for explosion-proof systems) and off-gas treatment evaluation as discussed in Chapter 7. The pilot study demonstrated the value of using vacuum-enhanced recovery in this type of application. A modular vacuum-enhanced recovery system was installed and operated for approximately $1^1/_2$ years, at which time concentrations were significantly reduced and efforts were underway to achieve regulatory approval for site closure.

TYPES OF SYSTEMS

There are basically two types of systems: a single-pump system, which uses a combined liquid/vapor pump, and a dual-pump system, which uses separate liquid and vapor pumps. The single-pump system uses a single-pump with a drop pipe to withdraw both fluids and vapors (Figure 14). Because we are dealing with a suction lift system, we are limited to a maximum theoretical lift of about 34 ft. In practice, this type of system usually is limited to lifts of 15 ft or so and thus is applicable for shallow aquifers only. The benefits of a single-pump system are lower capital and lower operational and maintenance costs. In addition, since

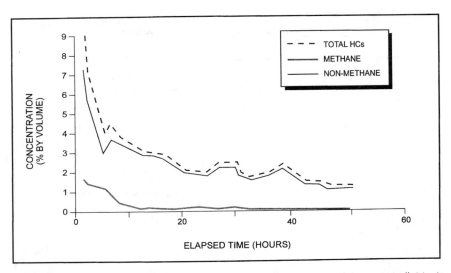

Figure 13 Hydrocarbon concentration in off-gases during vacuum enhancement pilot test.

all the equipment is above ground, it is easier to install and maintain. Presented in Figure 15 is a photograph showing a typical single-pump system that is pneumatic and operates off of compressed air available at the facility. Disadvantages of the suction lift system include limited depth, the problem of balancing vacuum at multiple wells, and the higher vacuum required to maintain the lift to produce fluids, which limits design flexibility.

Two-pump systems can be used for greater depths; for these systems, a vacuum is applied at land surface and a second downhole pump is used to withdraw fluids (Figure 16). These systems are easier to balance and operate when multiple extraction wells (greater than five) are involved and allow more design flexibility after selecting the optimum vacuum pressure. Shown in Figure 17 is a photograph of a two-pump mobile system that was used on the pilot studies described in a previous section. Disadvantages include the increased capital, operation, and maintenance costs of two pumping systems and the care that must be taken in selecting a downhole pump that will operate under vacuum.

If the pump is not designed properly (available net positive suction head is not greater than that required by the pump) many serious problems can result. There will be a marked reduction in head and capacity, or even a complete failure to operate. Problems also can include excessive vibration or erosion of the pump parts resulting in reduced life.

These systems can be very simple and inexpensive by using a small pneumatic diaphragm pump, for example, to extract both fluids and gases in a low permeability setting. The pump may be capable of moving only about 1 scfm of air; however, that can be quite effective, particularly if the goal is to maintain a good vacuum as opposed to maintaining high air flows. The systems also can be relatively complex, such as a liquid-ring vacuum-enhanced system that can be

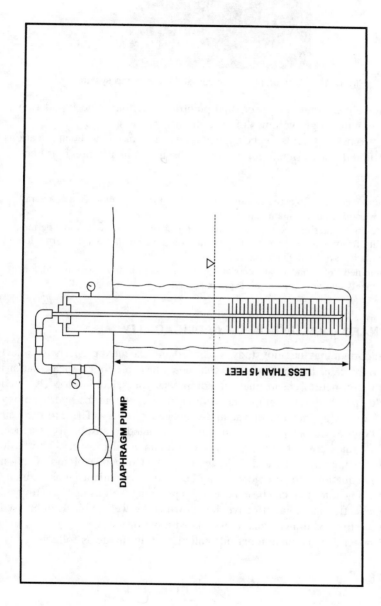

Figure 14 Schematic showing the configuration of a single pump vacuum-enhanced recovery system.

Figure 15 Photograph of a pneumatic single pump system.

modified for either single-pump or dual-pump applications. The liquid ring is capable of moving larger volumes of air at higher vacuums.

These two basic systems can be applied to the vacuum-enhanced applications to remove liquid and residual contaminants discussed in the previous section including:

- Liquid recovery: Objective is simply to increase the rate, influence, and overall recovery of liquid contaminants.
- Vacuum dewatering and vapor recovery: Objective is to dewater a low permeability formation and then use the vacuum to move air through the formation to volatilize or biodegrade the residuals.
- Combined liquid and vapor recovery: Objective is to both remove liquid contaminants and volatilize or biodegrade residuals.

PRELIMINARY EVALUATION OF APPLICABILITY

It is important that the pilot study is planned and executed properly to collect all of the right data. Before you even consider a test, preliminary calculations should be performed to ensure that the equipment you select is properly sized for the field application. It is important to first make sure that the hydrogeology is suitable for the application of vacuum-enhanced recovery. The expected air and water flows at various pressures should be estimated to ensure, for instance, that the flow rates are not too high for vacuum-enhanced recovery. The basic equation for estimating water and air flow rates is the Cooper/Jacob modification of the Theis equation shown below. It can be used to provide a quick estimate of the water flow rates that can be expected at a pumping well using the estimated vacuum application and total effective drawdown at the well. It also can be used to calculate estimated drawdowns in associated monitor wells.

The Cooper-Jacob equation permits calculating discharge as follows:

$$Q = \frac{s_t}{264 \log \frac{0.3Tt}{r^2 S}}$$

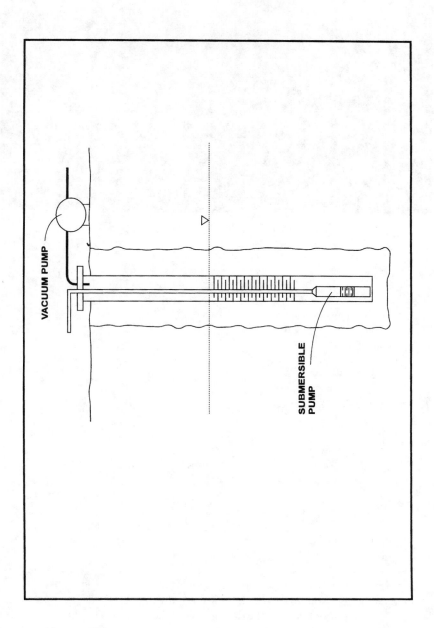

Figure 16 Schematic showing the configuration of a two-pump vacuum-enhanced recovery system.

Figure 17   Photo of a two-pump mobile system.

where

Q = discharge, in gpm
s_t = effective drawdown corrected for dewatering, in ft
T = transmissivity, in gpd/ft
t = pumping time, in days
r = borehole radius, in ft
S = storage coefficient

To correct for dewatering

$$s_t = s_a - \frac{s_a^2}{2b_w}$$

where

s_a = actual drawdown
b_w = aquifer thickness

This formula requires that you also have estimates of the site transmissivity, storage coefficient, and expected time of pumping.

Let us run through a typical example where the hydrogeology is suitable for consideration of a vacuum-enhanced recovery system. Assumptions are

Hydraulic conductivity $(K) = 10^{-4}$ cm/sec = 2.1 gpd/ft^2.
b_w = 20 ft for vadose zone.
b_g = 20 ft for saturated zone.
100% efficient well (both air and water).
$T = Kb_w$ = 42 gpd/ft.

At maximum drawdown

$$s_t = 20 - \frac{20^2}{(2)(20)} = 10 \text{ ft}$$

This is the "effective available drawdown" if no vacuum is applied.

Assume a liquid-ring pump will be used to apply a vacuum of 20 inHg. To get inches of water vacuum, multiply 20 by 13.55, yielding 271 inches of water, or 22.6 ft. This is the additional available drawdown applied to the well, bringing the total available drawdown to 22.6 plus 10, or 32.6 ft. At maximum drawdown, the water production with vacuum will be 3.26 times that without vacuum.

Using T = 42 gpd/ft, t = 30 days, S = 0.05, and r = 0.5 ft, the pumping rate (Q) is 0.355 gpm at 10 ft of drawdown (no vacuum); it increase to 1.16 gpm at 32.6 ft of drawdown (with 20 inHg vacuum). This preliminary result suggests that vacuum-enhanced recovery is suitable for this application, so the next step is to get a handle on expected air flow rates.

The air flow rates can be estimated in a similar manner using several formulas as follows.

Determine gas transmissivity, T_g. Gas conductivity is

$$K_g = K_w \frac{\rho_g \cdot \mu_w}{\rho_w \mu_g}$$

where

K_g = gas conductivity
K_w = hydraulic conductivity
ρ_g = gas density (0.0013 at 68°F)
ρ_w = water density (1 at 68°F)
μ_g = gas viscosity (183 μp at 68°F)
μ_w = water viscosity (10,000 μp at 68°F)

Thus,

$$K_g = 2.1 \; \frac{0.0013}{1} \cdot \frac{10,000}{183}$$

$$= 0.149 \text{ gpd/ft or } 0.02 \text{ ft/day}$$

Finally,

$$T_g = K_g b_g$$

$$T_g = 20 \cdot 0.02$$

$$= 0.4 \text{ ft}^2 / \text{day}$$

To estimate the air flow, we can use the Theis equation after correcting the drawdown (vacuum) for gas expansion. The gas expansion correction has the same form as the dewatering correction:

$$S_{eff} = S_a - \frac{s_a^2}{2P_{ATM}}$$

where

S_{eff} = effective vacuum used in Theis equation
S_a = actual vacuum
P_{ATM} = atmospheric pressure (405 inH$_2$O)

Since the vacuum is 271 inH$_2$O

$$S_{eff} = 271 - \frac{271^2}{2 \cdot 405}$$

$$= 180 \text{ inH}_2\text{O}$$

The Theis equation is

$$s = \frac{528Q}{T} \log \frac{R}{r}$$

where

s = vacuum, in ft of air
Q = discharge, in cfm
T = transmissivity, in ft^2/day

R = radius of influence, in ft
r = borehole radius, in ft

By assuming the expected vacuum at the well and estimating air permeability and the expected radius of influence; the extraction rate can be estimated. The radius of influence obviously is not known until the pilot studies are performed; however, the calculations are not highly sensitive to this parameter, thus using a value of 40 ft provides a good approximation. To express s in inches of water instead of ft of air, multiply by $12 \cdot 0.0013$:

$$s = \frac{8.23Q}{T} \log \frac{R}{t}$$

Solving for Q,

$$Q = \frac{sT}{8.23 \log \dfrac{R}{r}}$$

Using s = 180 inH$_2$O, T = 0.4 ft^2/day, R = 40 ft, and r = 0.5 ft,

$$Q = 4.6 \text{ cfm}$$

The preliminary calculations, as demonstrated in this example, indicate that conditions appear favorable for applying vacuum enhancement. As a next step, a pilot study needs to be performed. The pilot study needs to provide you with important data required for full-scale design.

PILOT TEST PROCEDURES

When the results of the detailed evaluation and comparative analysis indicate that vacuum-enhanced recovery may be applicable, then a pilot study should be conducted at the site to determine site specific parameters. Data from the pilot test program would be utilized to substantiate the preliminary calculations and to refine assumptions regarding such fundamental parameters as radius of influence, pumping rate, vacuum to be applied, and air and water quality. The collection and analysis of site-specific data would facilitate more accurate evaluation and final design of the remedial system. The pilot test plan would include the following:

- Installation of test wells
- Test method and monitoring
- Mass removal estimation

Test and Monitoring Wells

The test recovery well and monitoring wells should be installed for conducting the pilot test. The test recovery well should be installed in the vicinity of the impacted area of ground water and soils. The test recovery well should be screened both in the saturated and unsaturated area of the subsurface. It is recommended that at least four monitoring wells be installed at 10, 30, 50, and 100 ft away from the test recovery well. The wells should be designed and developed properly using a continuously wrapped well screen (No. 10 slot) and a fine grained sand pack.

Test Method

A vacuum pressure is applied to the test recovery well to evacuate the well and surrounding soils of liquids and air or volatile organic compounds. The lower section of the test and monitoring well casings are screened and the test well is equipped with a drop tube which will extend below the well static liquid level to near the bottom of the casing. The well casing head is sealed to withstand the applied vacuum pressure, and the vacuum is applied to the drop tube to evacuate the well.

A number of the monitoring wells are also equipped with drop tubes, which will extend below the well static liquid level to near the bottom of the casing, and the well casing heads are sealed. Well liquid levels are measured in the drop tube (with adjustments for casing side vacuum pressure). A pressure tap is provided on the casing to measure the induced soil vacuum pressure.

The portable vacuum pump is a centrifugal, liquid-ring type and is capable of producing and sustaining vacuum pressures from 0 to 24 inHg (from 0 to 27 ftH$_2$O). The liquid-ring pump is especially suited for this application because of its high vacuum pressure capability and because of the minimum risk for internal source of ignition while compressing potentially explosive mixtures of air or volatile organic compounds.

Monitoring

Prior to beginning the design test, the fluid levels in all of the on-site monitoring wells are measured, using an electronic water level indicator or electronic product/water interface probe. The recovery well should be surveyed and tied into the elevation network for the existing monitoring wells. During the course of the design test, all parameter measurements are recorded. Measured parameters, monitoring techniques, and time schedules include the following:

- Monitoring well fluid levels measured with electric fluid level indicators.
- Monitoring well casing vacuum pressure measured with calibrated vacuum pressure gauge or well manometer each hour.
- Liquid levels in other on-site wells will be observed during the test. Changes will be noted, and all values recorded.

- Test well-applied vacuum pressure measured with calibrated vacuum pressure gauge or mercury manometer each hour.
- Test well liquid production measured continually with totalizing-type turbine meter; accumulated total read and recorded hourly. Water tank will be gauged at the conclusion of the test to verify total production.
- Vapor volume of test recovery well volatile organic compounds continuously measured and recorded using an orifice meter and chart recorded.

This procedure is applicable for a single-pump system, which is suitable for depths less than 15 ft. Figure 18 shows the typical layout of the pilot test system for a vacuum-enhanced recovery. Note the placement of vacuum gauges and flow meters. A vacuum gauge is located at the recovery well and at the suction separator.

The most important aspect of the vacuum well-head is tight seals (Figure 19). These well-heads should be fabricated and tested prior to going into the field and should have a drop tube installed to measure fluid levels. The drop tube must be extended well below the fluid level and cannot be used if the vacuum is expected to exceed the submergence of the drop tube. This is only a consideration in close proximity to the extraction well. The use of a drop tube will prevent the measurement of floating hydrocarbons, if present. However, this is not critical since it is the total head we are interested in for design calculations.

Mass Removal Estimation

Vacuum-enhanced recovery will reduce the product mass by removing product in the free phase, dissolved phase, and vapor phase. The free product removed by pumping and the organic concentrations in ground water and vapor indicate the mass removed in the dissolved and vapor phases. The total mass removed during vacuum-enhanced recovery should be calculated based on the pilot test results. The free product mass is generally measured by calculating the amount removed from the oil/water separator, and the dissolved groundwater portion is calculated by multiplying the total volume of ground water pumped by the weighted average concentration of dissolved constituents measured in water samples collected during the test (three to five samples). Lastly, the mass in the air phase is calculated by multiplying the total volume of air discharged throughout the test by the weighted average concentration measured in the three to five air samples collected during the test.

In addition, soil vapor concentrations also should be monitored to assess the potential biological activity during the pilot test. The procedure to do this is described in detail in Chapter 3 in Section V.

SYSTEM DESIGN

After the pilot test is completed, there are a number of engineering design parameters that must be determined to design a full-scale system:

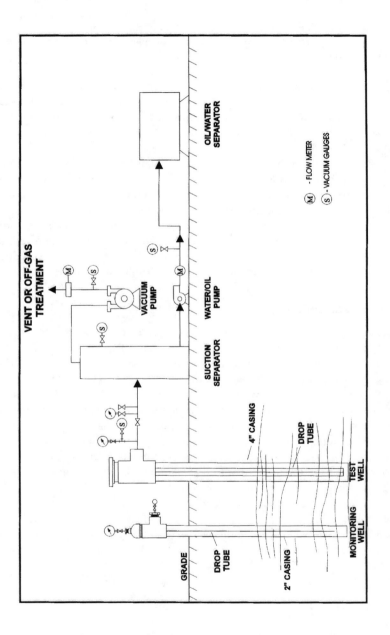

Figure 18 Schematic showing a typical layout of a pilot test system.

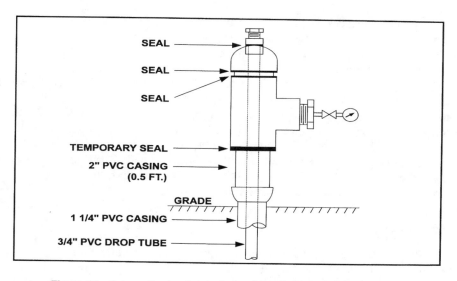

Figure 19 Schematic showing typical well-head details for monitor wells.

- Determine groundwater influence.
- Determine well spacing based on groundwater influence.
- Determine design flow rate.
- Select water treatment option (from groundwater concentrations).
- Determine required vacuum pressure.
- Determine venting influence.
- Determine air flow rate.
- Select off-gas treatment (from vapor concentrations).
- Evaluate the presence of biological activity.
- Select equipment.
- Estimate cleanup time.

The design is complicated by the need to evaluate both the groundwater and vacuum systems, and as such there is no "cookbook" approach that optimizes the design, but rather a detailed analysis of "tradeoffs" needs to be made to select the most suitable application for achieving the remedial goals. As a starting point, let us review the effect of vacuum application on liquid production, effective drawdown, and selection of an appropriate pumping level. The pumping level in the well is also a consideration in system design. Depending on the remedial objectives, you may want to maintain fluid levels in the well or maximize the drawdown, which will affect pump selection, system selection, and operation. Scenarios in which you may want to maximize fluid levels include situations where you are trying to use water as the carrier and thus want to maintain high fluid production rates.

On the other hand, situations in which you want to maximize drawdown would include those where you would like to dewater the aquifer and rely on air as the primary carrier to purge the dewatered zone of contaminants or to stimulate bioactivity in the dewatered zone. In this situation, the objective is to lower groundwater levels so that air can move through a larger portion of the contaminant zone; therefore, maintaining high air flow is more important than maintaining fluid production.

As we indicated in the beginning when you apply a vacuum of 10 ft of water, that amount is added to the available drawdown. If for example, you have 9 ft of available drawdown without a vacuum, then this converts to 19 ft of available drawdown when 10 ft of vacuum is applied. Consequently, the flow rate would essentially double using a vacuum-enhanced recovery system. During pump selection, you need to keep in mind that you have just added 10 ft to your total dynamic head (TDH) requirements. Also, the available net positive suction head (NPSH) should be greater than the NPSH required by the pump.

Well Design

The use of high vacuums coupled with low permeability formations can result in rapid well plugging or silting. The reduced pressures can result in more rapid precipitation of dissolved inorganic constituents on the well screen or gravel pack or within the formation. The increased gradients can result in fines plugging the screens or silting of the wells in poorly designed wells. Thus, proper well design and wrapped screens are normally employed, and periodic redevelopment of the wells may be necessary.

A thorough understanding of hydrogeologic conditions, aquifer conditions, and the most effective well development procedures is required for the completion of high-yielding, efficient recovery wells.

Listed below are some of the parameters that should be considered in designing recovery wells:

- Proper well design procedures for sizing gravel pack and screen
- Well diameter sufficient for the recovery equipment
- Appropriate slot size and open area to reduce plugging
- Screen length
- Depth to product/water from ground surface
- Product characteristics

A good reference book on designing wells including slot size and gravel paths and on proper well development techniques is *Groundwater and Wells* (Driscoll, 1986).

WELL SPACING

The first step in system design is to determine well spacing. The optimal well spacings will vary depending on your remedial objective. For instance, if your objective is to contain ground water and to prevent further movement downgradient, then the recovery wells would be spaced farther apart than if you were trying to achieve "maximum" dewatering so air can be the primary carrier.

In the latter case, the system would be designed with recovery wells much closer together and the principals set forth in Chapter 4 would be applied to achieve optimum air flow in the dewatered zone.

In either case, well spacings should first be determined based on groundwater influence. Information from the pilot test is used to determine the effective drawdown and flow rate which are affected by the vacuum level and associated pumping rate information, such as that shown in the graph in Figure 20, which depicts the effective drawdown vs. distance at various flow rates generated from a vacuum test. This information can be used to estimate well spacing and the influence of pumping on liquid levels. As a rule of thumb, do not consider the effective radius of influence beyond drawdown values of 0.10 ft. One important note concerning Figure 20 is that it shows increasing the vacuum does not necessarily increase the capture zone all that significantly. You get diminishing returns as vacuum pressure continues to increase.

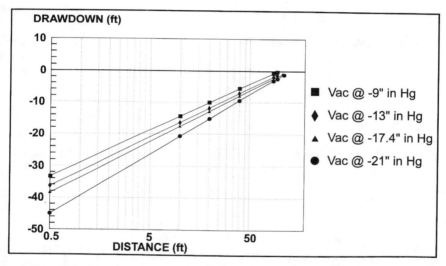

Figure 20 Distance vs. drawdown from a vacuum-enhanced recovery pilot test.

As a first cut in determining well spacings, use the effective radius of influence of the well at the expected vacuum pressure and plot overlapping cones to estimate the number and location of wells. This generally is a first-cut evaluation and leads to system overdesign because the radius of influence during the short-term pilot studies is nonequilibrium conditions. However, in certain cases (particularly in systems where you are trying to dewater and use air as a carrier) this approach may suffice.

For larger systems it is more appropriate to design the well spacings based on long-term operations when "apparent" equilibrium conditions are reached. To determine the capture zone under these conditions, the data collected during the pilot studies can be used with the approaches outlined below to evaluate capture zones under "apparent" equilibrium conditions and optimize well placement. It is important to note that use of conventional capture zone equations to calculate capture zones in tight sediments are not appropriate. Use of these equations results in prediction of capture zones that are unreasonably large. To demonstrate, let us look at an example of an "over-predicted" capture zone in a low-permeability formation. We will examine a hypothetical 20-ft thick silt formation having a hydraulic conductivity of 2.1 gpd/ft^2 (10^{-4} cm/sec), a storage coefficient of 0.05, and a hydraulic gradient of 0.01 ft/ft. Assume that a fully penetrating, 100% efficient extraction well operates for 30 consecutive days, resulting in a drawdown of 10 ft. Assume further that the borehole radius is 0.5 ft (12-in. borehole).

The discharge rate of the well may be calculated using the Cooper-Jacob equation, but the observed drawdown must be corrected for dewatering first. As stated earlier in this chapter, the dewatering correction is as follows:

$$s_t = s_a - \frac{s_a^2}{2bw}$$

With an observed drawdown of 10 ft,

$$s_t = 10 - \left(\frac{10^2}{2 \cdot 20} \right)$$

$$= 7.5 \text{ ft}$$

Similarly, using the Cooper-Jacob equation, the discharge is calculated as follows:

$$Q = \frac{s_t T}{264 \log\left(\frac{0.3Tt}{r^2 S} \right)}$$

The resulting discharge rate is 0.266 gpm.

The standard capture zone equations calculate the distance to the stagnation point (x_0), the capture width at the well (w_0), and the upgradient capture width (w) as follows (assuming consistent units):

$$x_0 = \frac{Q}{2\pi Tl}$$

$$w_0 = \frac{Q}{2Tl}$$

$$w = \frac{Q}{Tl}$$

For the example presented here, a flow rate of 383 gpd (0.266 gpm) gives the following results:

$$x_0 = 145 \text{ ft}$$

$$w_0 = 456 \text{ ft}$$

$$w = 912 \text{ ft}$$

Using vacuum-enhanced recovery techniques, the yield of this well would double or triple depending upon the amount of vacuum applied. Assuming we doubled the discharge rate to 766 (0.532 gpm), the capture zone dimensions compute to the following:

$$x_0 = 290 \text{ ft}$$

$$w_0 = 912 \text{ ft}$$

$$w = 1824 \text{ ft}$$

Based on this analysis, it is easy to see that remediation design engineers should not rely on a 0.5 gpm well to provide over one third of a mile of capture width. In practice, intermittent recharge events "swamp out" the cone of depression periodically because an ordinary recharge event can overwhelm the small discharge rates generally associated with tight formations. With the cone of depression being flooded out periodically, its lateral extent is limited and the well is not able to influence gradients at the great distances predicted by conventional capture theory.

To overcome the problems associated with conventional capture zone equations, an algorithm has been prepared that lead to a more conservative design. The essence of the procedure is to base capture on the configuration of the cone of depression after a fixed, limited pumping time — say, 30, 60, or 90 days. In other words, after the fixed time has passed, we assume we gain no further growth in the cone of depression. After selecting the arbitrary pumping time (perhaps 30 days for humid climates and 90 days for desert climates), drawdowns and gradients are calculated based upon the Theis equation (not the log equation) and capture analysis is based upon the resulting drawdown configuration. A description of this procedure follows, along with several required graphs and a few examples.

After the arbitrary pumping time has been chosen, the first step is to compute the so-called log-extrapolated radius of influence of the well, R. This is done primarily for mathematical convenience because R consolidates several other parameters. (R is commonly called the "radius of influence" because it is the distance to zero drawdown on an extrapolated semilog distance-drawdown graph. However, it is not a true radius of influence because the Theis equation predicts some additional drawdown beyond this point.) R may be computed from the following equation:

$$R = \sqrt{\frac{0.3\ Tt}{S}}$$

where

R = log-extrapolated radius of influence, in ft
T = transmissivity, in gpd/ft
t = pumping time, in days
S = storage coefficient

If you are using consistent units, i.e., transmissivity in ft²/day, the equation is a follows:

$$R = \sqrt{\frac{2.246\ Tt}{S}}$$

The log-extrapolated radius of influence is significant in that we will want to express capture zone dimensions in terms of R.

Differentiating and manipulating the Theis equation gives rise to the following expression relating Q, discharge, to x_0, distance to downgradient stagnation point.

$$Q = 2\ \pi\ Tix_0 e \left(\frac{x_0^2}{1,781\ R^2} \right)$$

This is the same as the conventional capture-zone equation except for the exponential term. When the exponent is small (x_0 is small in relation to R), the exponential term is close to 1 and the standard capture zone equations work just fine. As x_0 increases, however, to a significant fraction of R or beyond, the exponential term is substantially greater than 1 and the extraction well must produce more water than would be determined by conventional analysis. In essence, the exponential term represents a multiplier that must be applied to the Q computed from the standard equations.

Figure 21 shows the magnitude of the exponential term as a function of the ratio x_0/R. For example, if x_0 is half the radius of influence, Q will be 1.15 times the conventional calculation. If x_0 equals R, the discharge must be 1.75 times that calculated from conventional analysis. And if x_0 equals 2 R, nearly a 10-fold increase in discharge rate is required over conventional theory.

Conventional theory predicts that the width of the capture zone at the extraction well will be π times the distance to the stagnation point. That is,

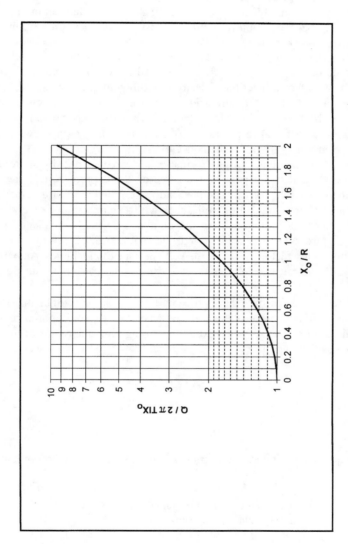

Figure 21   Ratio of required Q to conventional Q as a function of distance to stagnation point.

$$w_0 = \pi \, x_0$$

For tight formations (time-limited cones of depression), however, the ratio w_0/x_0 decreases as discharge rate and capture zone size increase. Figure 22 shows how this ratio decreases with increasing x_0/R. Figure 22 was developed empirically by using analytic element modeling and particle tracking to assess the relationship between capture width and distance to stagnation point.

Figure 21 was based upon an exact equation, whereas Figure 22 was determined empirically. Combining these two graphs produces Figure 23, which shows the magnitude of the discharge increase required for capture compared to conventional theory, all as a function of desired capture width. For example, reading from the graph, when the capture width is twice the radius of influence, the discharge must be 1.4 times the value calculated using the conventional equations. For a capture width equal to the radius of influence, the graph shows that the required discharge is 1.09 times the conventional discharge.

Figure 23 shows that in tight formations we pay a penalty in terms of discharge rate and that the penalty increases as the capture zone dimensions approach and exceed the log-extrapolated radius of influence. If the remediation design includes a large number of wells, each with a small capture zone, the total required discharge is minimized. As the number of wells is reduced and the individual capture zone size is increased, the required discharge rate increases by the multiplication factor shown on the vertical scale of the figures. An optimum remediation design must weigh the costs of drilling more wells versus pumping more water.

An example will illustrate these procedures. Calculations will be made based upon capture width, w_0, because this is the parameter of greater importance. Using the 20-ft silt described above and a 30-day pumping time, the radius of influence is calculated as follows:

$$R = \sqrt{\frac{0.3 \cdot 42 \cdot 30}{0.05}}$$

$$= 87 \text{ ft}$$

The required flow rate for a particular capture width may be computed as follows:

1. Select the desired capture width: for example, a w_0 of 200 ft.
2. Compute discharge required using conventional equation:

$$Q = 2TIw_0$$

$$= 2 \cdot 42 \cdot 0.01 \cdot 200$$

$$= 168 \text{ gpd } (0.12 \text{ gpm})$$

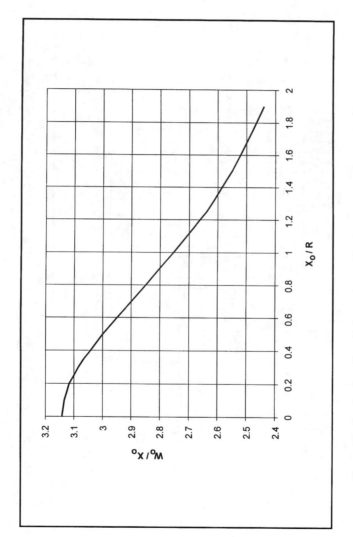

Figure 22 Decline in w_0/x_0 with increasing x_0 (R = log-extrapolated radius of influence).

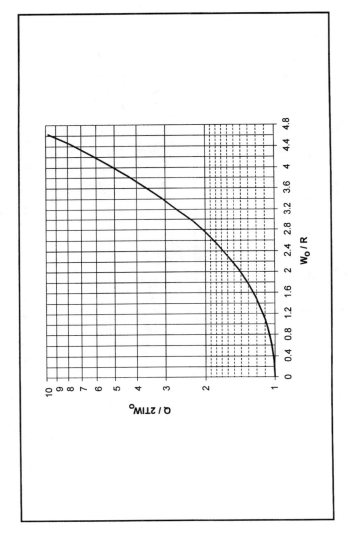

Figure 23   Ratio of required Q to conventional Q as a function of desired capture width.

3. Calculate w_0/R and use Figure 23 to determine the multiplier to apply to the conventional rate. The ratio w_0/R equals 2.3 and, reading from the graph, the multiplier is 1.6. Thus, the required Q is

$$Q = 168 \cdot 1.6$$

$$= 269 \text{ gpd } (0.19 \text{ gpm})$$

Repeating this process for another assumed value of w_0 — for example, 300 ft — yields the following:

$$Q = 2.42 \cdot 0.01 \cdot 300 \quad (\text{conventional equation})$$

$$= 252 \text{ gpd } (0.187 \text{ gpm})$$

$$w_0 / R = 3.45$$

$$\text{Multiplier} = 3.2 \text{ (from Figure 23)}$$

$$Q \text{ actually required } = 252 \cdot 3.2$$

$$= 806 \text{ gpd } (0.56 \text{ gpm})$$

These same calculations were performed for capture widths ranging from 50 ft to 400 ft in 50-ft increments. Table 1 shows the results, including the discharge associated with the conventional approach and the actual required discharge associated with this procedure. Recall that earlier analysis showed a required discharge rate of 0.266 to achieve a w_0 of 456 ft. According to Table 1, this discharge rate would result in a capture width between 200 and 250 ft. Similarly, earlier calculations showed that a discharge rate of a little over 0.5 gpm resulted in more than 900 ft of capture, whereas Table 1 predicts less than 300 ft of capture.

Table 1   Comparison of Conventionally Calculated Discharge Rates and Actual Required Rates for 30 Days of Uninterrupted Pumping

w_0 (ft)	w_0/R	Q from conventional equation (ft³/day)	(gpm)	Multiplier from Figure 23	Q actually required (ft³/day)	(gpm)
50	0.58	42	0.029	1.02	43	0.030
100	1.15	84	0.058	1.1	92	0.064
150	1.73	126	0.088	1.29	163	0.113
200	2.30	168	0.117	1.6	269	0.187
250	2.88	210	0.146	2.2	462	0.321
300	3.45	252	0.175	3.2	806	0.560
350	4.03	294	0.204	5.1	1499	1.041
400	4.60	336	0.233	9.5	3192	2.217

The calculations described here can be made more conservatively by choosing a shorter pumping time. For example, if a pumping time of 10 days is used in the above example, the radius of influence, R, is 50 ft. Recalculating the capture zone information produces the results shown in Table 2. Note that, in this case, a discharge rate of over 0.5 gpm would produce less than 200 ft of capture width, and a discharge rate about 0.25 gpm would produce a little over 150 ft of capture width.

Table 2   Comparison of Conventionally Calculated Discharge Rates and Actual Required Rates for 10 Days of Uninterrupted Pumping

		Q from conventional equation			Q actually required	
w_0 (ft)	w_0/R	(ft³/day)	(gpm)	Multiplier from Figure 23	(ft³/day)	(gpm)
50	1.00	42	0.029	1.09	46	0.032
75	1.49	63	0.044	1.2	76	0.053
100	1.99	84	0.058	1.4	118	0.082
125	2.49	105	0.073	1.75	184	0.128
150	2.99	126	0.088	2.35	296	0.206
175	3.49	147	0.102	3.35	492	0.342
200	3.98	168	0.117	5	840	0.583
225	4.48	189	0.131	8.2	1550	1.076

When using the formulas presented previously, it is important to incorporate well efficiency into these formulas. In our example problem shown, the well efficiency was assumed to be 100%; however, under field conditions associated with the types of sediments where vacuum-enhanced recovery is applied, the well efficiency can be substantially less than 100%. The pilot test will provide an insight into the expected well efficiency; however, it is important to keep in mind that well efficiency may vary in accordance with geologic changes across a site. Generally, as the grain size of the sediments opposite the well screen decrease, the well efficiency also decreases.

The well efficiency from a pilot test can be estimated by using the distance-drawdown graph (Figure 20). This is done by extending the straight line representing the profile of the cone of depression to show drawdown in the aquifer just outside the well. The theoretical drawdown represented by this point is for a 100% efficient well. (In a filter-packed well, the radius is taken from the center of the well to the outside of the filter pack). The well efficiency is calculated by dividing the theoretical drawdown in the well by the actual drawdown (Driscoll, 1986).

These methodologies will only give an approximation, and the size and complexity of the job will determine the extent of the hydrogeologic analysis which is required. In some instances, you are dealing with multiple tests of varying values as a result of variations in hydrogeology. If these variations in hydrogeology can be defined, then they must be taken into account in determining well spacing and system influence. If not, then a conservative estimate must be made.

Fluid Flow Rate

Based on well spacing and a hydrogeologic evaluation, the design fluid flow rate can be determined. The design flow rate will be a summation of the flows from the individual extraction wells and is actually developed hand-in-hand with well spacing. If the capture zones from individual wells overlap one another, then fluid flow from individual extraction wells will be less than if the capture zones merely "touch" each other. Thus, if the system design is based on plotting capture zones from individual wells with data collected from pilot studies, then the calculated fluid recovery rate will be higher than will actually occur, so this should be kept in mind. Of course, using the capture zone equations discussed in the previous section is generally the preferred method; however, often times data is not sufficient to get this sophisticated so using simpler methodologies and good engineering judgment may suffice. Either way, a well-run pilot study is the key to collecting good site specific data, and if you have limited funds, the money is better spent basing system design on this in lieu of sophisticated modeling approaches that are performed without the aid of site specific pilot tests.

Vacuum Pressure

Vacuum pressure selection can be determined by a series of curves that are generated from the pilot testing. Ideally, your optimum vacuum is the lowest vacuum that will give you the water production and air influence you need. However, other criteria may control such as suction lift or water flow requirements. You must balance your desired water discharge, air influence, and air flow to arrive at the optimum vacuum. The data collected from the pilot studies should be evaluated in a manner similar to that shown in Figure 24. Typically the vacuum decreases with time. You should anticipate that a decrease of 4 to 6 inHg at the extraction well will occur within 6 months of operation. You should carefully consider this during the design of the vacuum pump. The optimum vacuum is defined as the point at which higher vacuums result in diminishing benefits for water discharge, air influence, and air flow. This may drive your pumping system selection and the cost of increased vacuum would be weighed against the cost of a two-pump system.

A higher vacuum will result in a higher air flow rate and venting influence, greater fluid recovery rates, and greater groundwater influence. However, this is balanced by the increased cost and energy requirements of increasing the vacuum, and the increase is not linear. As the vacuum continues to increase, the relative increase in water and air flow rates declines as depicted in Figure 24, which shows a comparison of the pilot test data at various vacuum pressures vs. air flow rate and influence.

The selection of the vacuum pumping system also will influence the design vacuum pressure. The design pressure may be reduced below "optimum levels" if the use of a liquid-ring application can be avoided in favor of a less expensive alternative at a lower vacuum pressure, without significantly sacrificing effectiveness. This would be a function of a capital cost evaluation and operating and

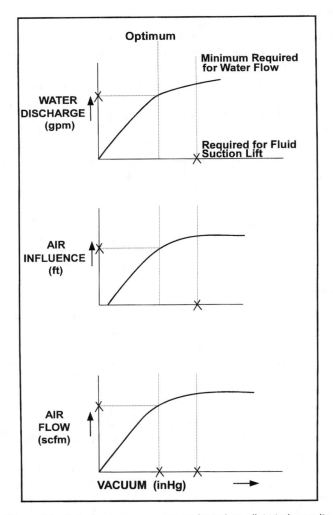

Figure 24 Selecting vacuum pressure based on pilot study results.

maintenance considerations. The final factor to evaluate is the type of pumping system selected. If a single-pump vacuum system is used, the minimum vacuum will be defined by the suction lift required to produce fluids from the well. Evaluation of the other factors may indicate that a two-pump type system is more cost effective since a lower vacuum pressure can be selected. Once the vacuum pressure has been selected, then you must determine the venting influence. In some applications, the use of air as the carrier for removal of residuals through volatilization or enhancement of bioactivity will be one of the primary applications of the vacuum system. A general rule of thumb is using a value of 0.1 inH$_2$O vacuum as the limiting effective vapor extraction influence. In most cases, the dominant issue for well spacing is the groundwater influence and not the venting influence. However, there may be some instances where the venting influence drives the selection of well spacing. This will be a function of the objectives of your remedial program.

Air-Flow Rate

The vacuum pressure selected above will set the air flow rate for the system. The application of the pilot test results to the remainder of the site requires a hydrogeologic evaluation as the consistency and/or variation of materials at the site, similar to that done for the groundwater flow estimation. If there are well-defined areas or known variations in site permeability, it is wise to run several short-term pilot tests (a few hours) to get a range of design parameters and flow rates. These can be used to estimate the variability in site-wide system production.

Off-Gas Treatment

An off-gas treatment system can be the most costly portion of the vacuum-enhanced treatment system, so care must be exercised to assure that the most cost effective approach is used and that changes in off-gas concentrations vs. time, which can occur quite rapidly, are understood. One of the primary concerns with the use of vacuum-enhanced systems at petroleum hydrocarbon sites is the high concentrations of hydrocarbons in the off-gas during the early phases of the remediation. These concentrations in the "several percent range" must be factored into the design selection. The high concentrations usually require the use of some form of thermal destruction, and explosion-proof equipment is needed in many cases. If off-gas concentrations are expected to decrease rapidly, you may want to switch technologies to reduce life cycle costs as discussed in Chapter 2.

Equipment Selection

All of the factors evaluated above will play a role in the selection of the equipment for a particular vacuum-enhanced option. Two of the factors will be discussed below, but a cost-benefit analysis needs to be performed that evaluates all of the parameters before a final selection is made.

A selection of the design vacuum level will dictate to some extent the type of vacuum system that must be employed at the site. At high vacuums (greater than 15 inHg or 19 ftH$_2$O) liquid-ring vacuum pumps (Figure 17) are normally employed; in the lower ranges (less than 8 inHg or 10 ftH$_2$O) regenerative blowers or positive displacement blowers (Figure 25) are normally employed; in the mid-range (between 8 and 15 inHg) lobe pumps (Figure 26) are used. These are strictly general ranges, and they vary widely depending upon the site conditions and the needs at the site. For instance, regenerative blowers could be put in series to increase their vacuum production rather than going to a lobe pump, as one example variation.

Figure 25   Regenerative blower.

For petroleum hydrocarbon remedial programs, another factor to consider is the need for explosion-proof systems due the high hydrocarbon vapor concentrations in the off-gas. This should be one of the primary factors considered in the system design and will impact the pump selection and system configuration.

MASS REMOVAL AND REACHING CLEANUP GOALS

When applied to the right compounds and in a favorable geologic setting, vacuum-enhanced recovery can be very effective in both mass removal and achieving cleanup goals. Its ability to achieve these goals varies depending on the application and contaminant involved, and in all cases, if the permeability is too low throughout the saturated thickness, then vacuum-enhanced systems are not applicable unless the geology is manipulated (Chapter 8) to make it suitable for use of a vacuum-enhanced system.

Enhanced Effectiveness — LNAPLs

Vacuum-enhanced recovery systems can be very effective in mass removal of LNAPLs, since applying a vacuum adds a second force (pressure differential) to fluid removal and can significantly extend the capture zone. It is very effective in fairly rapid recovery of LNAPLs, and the use as air as a carrier is much more effective in removing volatile organics from the dewatered zone than using water as a carrier itself. Since 70 to 90% of mass at sites containing LNAPLs are often in the non-aqueous phase, use of a vacuum-enhanced system for mass removal can be very beneficial. In regards to reaching cleanup goals, if the goal is removal

Figure 26 Lobe pump.

of LNAPLs from the subsurface, this technology can be very effective in achieving this goal, usually in a reasonable time frame of less than one year to several years, depending on site conditions.

Enhanced Effectiveness — Dissolved Phase

We have discussed how vacuum-enhanced systems are very effective in increasing the size of capture zones and in increasing water production rates. However, if the remedial goal is to bring groundwater cleanup to low constituent levels, such as drinking water standards, then vapor-enhanced recovery suffers from the same limitations as pump-and-treat programs. Achieving cleanup at low levels is difficult to achieve because although air is a good carrier of volatile organic compounds it can be applied only to the zone above groundwater levels (under pumping conditions). Thus, it may reduce contaminants to acceptable levels in this zone within reasonable time frames, but in the zone below the water table where water is the only carrier you still have the limitations associated with conventional pumping systems. Moreover, the lower permeable deposits often contain highly adsorptive materials which makes reaching such cleanup goals even more difficult than if the LNAPLs were located in a sandy environment where there is less adsorption to the earth materials.

One thing to keep in mind in setting cleanup goals at sites suitable for vacuum-enhanced systems is that these are always used in lower permeability environments which generally are not used for drinking water purposes. As such, the cleanup goals should reflect higher concentrations of contaminants than if treating a drinking water aquifer and should be set at levels that will protect the receiving body whether it is surface water or deeper aquifers. Realistic cleanup goals can be very important in shaving years off operating vacuum-enhanced recovery systems and should not be overlooked.

Enhanced Recovery — Dewatering

Use of vacuum-enhanced systems for dewatering and mass removal via vapor extraction can be very effective particularly in perched zones containing alternating lenses of sand and clays. Where this differs from the situation discussed above is that groundwater levels opposite the sand lenses, where most of the contaminants reside (because of the higher permeabilities), can be drawn below the sand lens, exposing the primary contaminant zone to remediation using air as the carrier. This is true where LNAPLs are present or if the contaminants are in the dissolved phase. Thus, not only do you achieve good mass removal, but it also becomes easier to achieve cleanup goals in more reasonable time frames because the sands are less adsorptive than the lower permeable clays, so air can more effectively remove volatile organic compounds.

REFERENCES

Blake, S.B. and Gates, M.M., "Vacuum Enhanced Recovery: A Case Study." Proc. Petroleum Hydrocarbons and Organic Chemicals in Groundwater: Prevention, Detection and Restoration. National Well Water Assoc./American Petroleum Inst., Houston, TX, 1986.

Blake, S.B. and Hall, R.A., "Monitoring Petroleum Spills with Wells: Some Problems and Solutions," Proc. Fourth National Symposium on Aquifer Restoration and Ground Water Monitoring, *NWWA*, pp. 305–310, 1984.

Cooper, H.H., Jr. and Jacob, C.E., "A Generalized Graphical Method for Evaluating Formation Constants and Summarizing Well Field History." *Trans. Am. Geophysical Union*, vol. 27, no. 4, 1946.

Driscoll, F.G., *Groundwater and Wells*, 2nd Ed., St. Paul, MN: Johnson Filtration Systems, Inc. 1986.

Powers, J.P., *Construction Dewatering: A Guide to Theory and Practice*, New York: John Wiley & Sons, 1981.

6

IN SITU AIR SPARGING

Suthan S. Suthersan

INTRODUCTION

In situ air sparging is a remediation technique that has been used since about 1985, with varying success, for the remediation of volatile organic compounds (VOCs) dissolved in the ground water, sorbed to the saturated zone soils, and trapped in soil pores of the saturated zone. This technology is often used in conjunction with vacuum extraction systems (Figure 1) to remove the stripped contaminants and has broad appeal due to its projected low costs relative to conventional approaches.

The difficulties encountered in modeling and monitoring the multiphase air sparging process (i.e., air injection into water saturated conditions) have contributed to the current uncertainties regarding process(es) responsible for removing the contaminants from the saturated zone. Engineering design of these systems, even today, is largely dependent on empirical knowledge. At this point, the air sparging process should be treated as a rapidly evolving technology with a need for continuous refinement of optimal system design and mass transfer efficiencies. The mass transfer mechanisms during *in situ* air sparging rely on the interactions among complex physical, chemical, and microbial processes, many of which are not well understood.

A typical air sparging system has one or more subsurface points through which air is injected into the saturated zone. It was commonly perceived for a long time (Angell, 1992; Sellers and Schweiber, 1992) that the injected air travels up through the saturated zone in the form of air bubbles; however, it is more likely that the air travels in the form of continuous air channels (Johnson et al., 1993; Wei et al., 1993; Ardito and Billings, 1990). The air-flow path will be influenced by pressure and flow of injected air and depth of injection, but mainly by structuring and stratification of the saturated zone soils (Johnson et al., 1993; Wei et al., 1993; Ardito and Billings, 1990). Significant channeling may result from relatively subtle permeability changes, and the degree of channeling will

0-87371-995-6/96/$0.00+$.50
© 1996 by CRC Press, Inc.

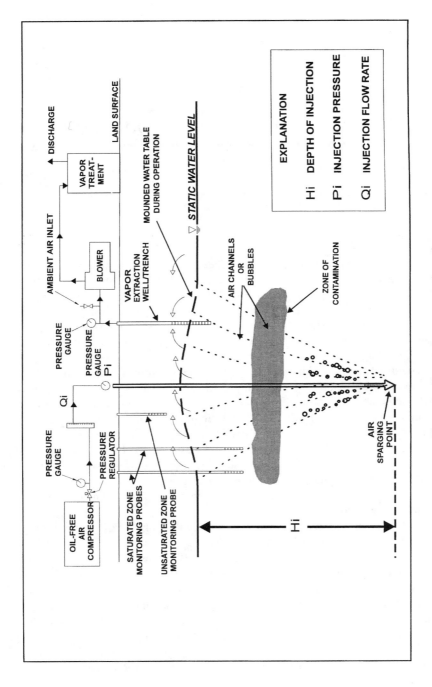

Figure 1 Air sparging process schematic.

increase as the size of the pore throats get smaller. Research (Wei et al., 1993) shows that even minor differences in permeability due to stratification can impact the sparging effectiveness.

In addition to conventional air sparging (injection of air is shown in Figure 1), many modifications of the techniques to overcome the geologic/hydrogeologic limitations will also be discussed in this chapter.

GOVERNING PHENOMENA

In situ air sparging is potentially applicable when volatile or easily aerobically biodegradable organic contaminants are present in water-saturated zones, under relatively permeable conditions. The *in situ* air sparging process can be defined as injection of compressed air at controlled pressures and volumes into water-saturated soils. The three main contaminant-removal mechanisms that occur during the operation of air sparging systems include: (1) *in situ* stripping of dissolved VOCs, (2) volatilization of trapped and sorbed phase contamination present below the water table and in the capillary fringe, and (3) aerobic biodegradation of both dissolved and sorbed phase contaminants resulting from delivery of O_2.

In Situ Air Stripping

Among the above contaminant-removal mechanisms, "*in situ* air stripping" may be the dominant process for some dissolved contaminants. The strippability of any contaminant is a function of its Henry's Law constant (vapor pressure/solubility). Compounds such as benzene, toluene, xylene, ethyl benzene, trichloroethylene, and tetrachloroethylene are considered to be very easily strippable (see Chapter 1 for a discussion of Henry's Law constants). During air sparging, dissolved compounds are transferred into the vapor phase and will be captured by a vapor extraction system (VES) once they migrate into the vadose zone.

However, a basic assumption made in analyzing the "air stripping" phenomenon during air sparging is that Henry's Law applies to the volatile contaminants and that all the contaminated water is encountered by the injected air. In-depth evaluation of these assumptions exposes the shortcomings and complexities of interphase mass transfer during air sparging.

First of all, Henry's Law is valid only when partitioning of mass has reached equilibrium at the air/water interface. However, the residence time of air, traveling in discrete channels, may be too short to achieve the equilibrium due to the high air velocities and short travel paths. Another issue is the validity of the assumption that the contaminant concentration at the air/water interface is the same as in the bulk water mass. The conventional application of air-to-water ratios in air stripping may not be applicable during *in situ* air sparging due to the above reasons.

Due to the removal of contaminants in the immediate vicinity of the air channels, it is safer to assume that the contaminant concentration is going to be lower around the channels. To replenish the mass lost from the water around the air channel, mass transfer by diffusion and convection must occur from water away from the air channels. Hence, it is perceivable that the density of air channels will play a significant role in mass transfer efficiencies by minimizing the distances required for a contaminant "molecule" to encounter an air channel. This limitation may also prevent this technology from reaching final cleanup criteria as discussed in Chapters 1 and 2.

On the positive side, *in situ* air sparging may also help to increase the rate of dissolution of the sorbed phase contamination and eventual stripping below the water table. This is due to the enhanced dissolution caused by increased mixing and the higher concentration gradient between the sorbed and dissolved phases under sparging conditions.

Direct Volatilization

The primary mass removal mechanism for VOCs present in the saturated zone during pump-and-treat operations is resolubilization into the aqueous phase and the eventual removal with the extracted groundwater. During *in situ* air sparging, direct volatilization of the sorbed and trapped contaminants is enhanced in the zones where air flow takes place. The volatile compounds do not have to transfer through the water to reach the air. If an air channel intersects pure compound, direct volatilization can occur. Direct volatilization of any compound is governed by its vapor pressure. Most volatile organic compounds are easily removed through volatilization.

In areas where significant levels of residual contamination of VOCs or non-aqueous phase liquids (NAPLs) are present in the saturated zone, direct volatilization into the vapor phase may become the dominant mechanism for mass removal. The high level of mass that the air can carry combined with the fast exchange of pore volumes results in a process that can remove significant pounds of contaminants in a relatively short period of time.

Biodegradation

In most natural situations, aerobic biodegradation of biodegradable compounds in the saturated zone is limited by the availability of oxygen. Biodegradability of any compound under aerobic conditions is dependent on its chemical structure and the environmental parameters such as pH and temperature. Some of the VOCs are considered to be easily biodegradable under aerobic conditions (e.g., benzene, toluene, acetone, etc.) and some of them are not (e.g., trichloroethylene and tetrachloroethylene).

Typical dissolved oxygen (DO) concentrations in uncontaminated ground water are less than 4.0 milligrams per liter (mg/L) and, under anaerobic conditions induced by the natural degradation of the contaminants, are often less than 0.5 mg/L. Dissolved oxygen levels can be raised by air sparging up to 6 to 10 mg/L under equilibrium conditions. This potential increase in the DO levels will contribute to enhanced rates of aerobic biodegradation in the saturated zone. This method of introducing oxygen to increase the dissolved oxygen levels to enhance biodegradation rates is one of the inherent advantages of *in situ* air sparging. The specific costs and methodology were discussed in Chapter 3, *In Situ* Bioremediation.

APPLICABILITY

Examples of Contaminant Applicability

Based on the discussion in the previous section, Table 1 describes the applicability of a few selected contaminants. In order for air sparging to be effective, the VOCs must transfer from the groundwater or from the saturated zone into the injected air, and oxygen present in the injected air must transfer into the ground water to stimulate biodegradation.

Table 1   A Few Examples of Contaminant Applicability for *In Situ* Air Sparging

Contaminant	Strippability[a]	Volatility[b]	Aerobic biodegradability[c,d]
Benzene	High (H = 5.5 x 10^{-3})	High (VP = 95.2)	High (t$_{1/2}$ = 240)
Toluene	High (H = 6.6 x 10^{-3})	High (VP = 28.4)	High (t$_{1/2}$ = 168)
Xylenes	High (H = 5.1 x 10^{-3})	High (VP = 6.6)	High (t$_{1/2}$ = 336)
Ethyl benzene	High (H = 8.7 x 10^{-3})	High (VP = 9.5)	High (t$_{1/2}$ = 144)
Trichloroethlene (TCE)	High (H = 10.0 x 10^{-3})	High (VP = 60)	Very low (t$_{1/2}$ = 7704)
Perchloroethlene (PCE)	High (H = 8.3 x 10^{-3})	High (VP = 14.3)	Very low (t$_{1/2}$ = 8640)
Gasoline constituents	High	High	High
Fuel oil constituents	Low	Very low	Moderate

[a] H = Henry's Law constant (atm-m^3/mol).
[b] VP = Vapor pressure (mmHg) at 20°C.
[c] t$_{1/2}$ = Half-life during aerobic biodegradation, hours.
[d] It should be noted that the half-lives can be very dependent on the site-specific subsurface environmental conditions.

In practice, the criterion for defining strippability is based on the Henry's Law constant being greater than 1×10^{-5} atm-m^3/mol. In general, compounds with a vapor pressure greater than 0.5 to 1.0 mmHg can be volatilized easily; however, the degree of volatilization is limited by the flow rate of air. The half-lives presented in Table 1 are estimates in ground water under natural conditions without any enhancements to improve the rate of degradation (enhancements are discussed in Chapter 10).

The constituents present in heavier petroleum products such as No. 6 fuel oil will not be amenable to either stripping or volatilization (Figure 2). Hence, the primary mode of remediation, if successful, will be due to aerobic biodegradation. Required air injection rates under such conditions will be influenced only by the requirement to introduce sufficient oxygen into the saturated zone.

Figure 2 qualitatively describes different mass removal phenomena in a simplified version under optimum field conditions. The amounts of mass removed by stripping and volatilization have been grouped together, due to the difficulty

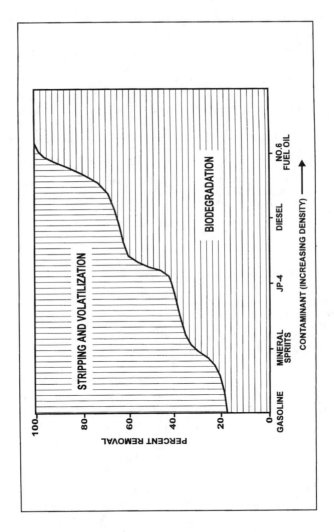

Figure 2   Qualitative presentation of potential air sparging mass removal for petroleum compounds.

in separating them in a meaningful manner. However, the emphasis should be placed on total mass removal, particularly of mobile volatile constituents, and closure of the site regardless of the mass transfer mechanisms.

Geological Considerations

Physical implementation of *in situ* air sparging is greatly influenced by the ability to achieve significant air distribution within the target zone. Good vertical pneumatic conductivity is essential to avoid bypassing or channeling of injected air horizontally, away from the sparge point. It is not an easy task to evaluate the pneumatic conductivities in the horizontal and vertical direction for every site considered for *in situ* air sparging.

Geologic characteristics of a site are very important when considering the applicability of *in situ* air sparging. The most important geologic characteristic is stratigraphic homogeneity or heterogeneity. The presence of lower permeability layers under stratified geologic conditions will impede the vertical passage of injected air. Under such conditions, injected air may accumulate below the lower permeability layers and will travel in a horizontal direction, thus potentially enlarging the contaminant plume (Figure 3). Any obvious high-permeability layers will also cause the air to preferentially travel laterally, thus potentially causing an enlargement of the plume (Figure 3). Horizontal migration of injected air limits the volume of soils that can be treated by direct volatilization and can cause safety hazards if hydrocarbon vapors migrate into confined spaces such as basements and utilities. Hence, homogeneous geologic conditions are essential for the success and safety of *in situ* air sparging.

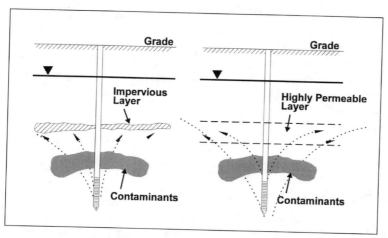

Figure 3 Potential situations for the enlargement of a contaminant plume during air sparging.

Both vertical pneumatic conductivity and the ratio of horizontal-to-vertical permeability increase with decreasing average particle size of the sediments in the saturated zone. The reduction of vertical permeability is directly proportional to the effective porosity and average grain size of the sediments (Bohler, et al., 1990). Hence, based on the empirical information available, it is recommended that application of in situ air sparging be limited to saturated zone conditions where the hydraulic conductivities are greater than 10^{-3} cm/sec (EPA, 68-03-3409, 1992).

It may not be possible to encounter homogeneous geologic conditions across the entire cross section at most sites. Hence, the optimum geologic conditions for air sparging may be where the permeability increases with increasing elevation above the point of air injection. Decreasing permeabilities with elevation above the point of air injection will have the potential to enlarge the plume due to lateral movement of injected air.

DESCRIPTION OF THE PROCESS

Air Injection Into Water-Saturated Soils

The ability to predict the performance of air sparging systems is limited by the current understanding of air flow in the water-saturated zone and limited performance data. There are two schools of thought in the literature describing this phenomenon. The first, and the widely accepted one, describes the injected air as traveling in the vertical direction in the form of discreet air channels. The second, describes the injected air as traveling in the form of air bubbles. Air flow mechanisms cannot be directly observed in the field; however, conclusions can be reached by circumstantial evidence collected at various sites and laboratory-scale visualization studies.

Sandbox model studies (Wei et al., 1993; Johnson, 1995) tend to favor the "air channels" concept over the "air bubbles" concept. In laboratory studies simulating sandy aquifers (grain sizes of 0.75 mm or less) at low air injection rates, stable air channels were established in the medium at low injection rates, whereas under conditions simulating coarse gravel (grain sizes of 4 mm or larger) the injected air rose in the form of bubbles. At high air-injection rates in sandy, shallow, water-table aquifers, the possibility for fluidization (loss of soil cohesion) around the point of injection exists (Johnson et al., 1993; Johnson, 1995; Lundegard and Anderson, 1993), and thus the loss of control of the injected air may arise.

Mounding of Water Table

As the injected air enters the saturated zone, the water-table elevation adjacent to the sparge point may rise due to the displacement of pore water by injected air. Displacement of ground water may initially form a mound around the injection well, although there is some evidence in the literature that this phenomenon is very transient (Johnson et al., 1993; Johnson, 1995; Lundegard and Anderson, 1993; Ardito and Billings, 1990). Some concerns have been raised by regulatory agencies regarding the potential for enhanced transport of dissolved contaminants caused by the movement of ground water away from the induced mound. However, the mound would not have the same kinetic energy as a pure water mound, and no evidence has been found that supports the spreading supposition.

Distribution of Air Flow Pathways

It is often envisioned that air flow pathways developed during air sparging form an inverted cone with the point of injection being the apex. This would be more true if soils were perfectly homogeneous or of larger sediments and injected air flow rate was low. Laboratory experiments simulating mesoscale heterogeneities in soil particle sizes resulted in distorted cone shapes caused by channels expanding, coalescing, and migrating upwards (Wei et al., 1993). Thus, it is reasonable to expect that distorted air channels will predominate in natural settings. During laboratory experiments using homogeneous media with uniform grain sizes, symmetric air flow patterns about the vertical axis were observed (Wei et al., 1993). However, media formed with mixed grain sizes yielded nonsymmetric air flow patterns. The asymmetry apparently resulted from minor variations in the permeability and capillary air entry resistance which resulted from pore-scale heterogeneity. Hence, under natural conditions, it is realistic to expect that symmetric air distribution will never occur. These same experiments also indicated that the channel density increased with increased air flow rates, since higher volumes of air occupy an increased number of air channels.

It is reported in some literature (Brown, 1992a) that, at low sparge pressures, air travels 1 to 2 ft horizontally for every foot of vertical travel. However, it has to be noted that this correlation was not widely observed. It was also reported that as the sparge pressure is increased, the degree of horizontal travel increases (Johnson, 1995; Lundegard and Andersen, 1993; Brown, 1992a). Visual observations have indicated that air flow channels extend 10 to 20 ft away from the air-injection point, independent of flow rate and depth of sparge point (Bohler et al., 1990).

SYSTEM DESIGN PARAMETERS

In the absence of any reliable models for the *in situ* air sparging process, empirical approaches are used in the system design process. The parameters of significant importance in designing an *in situ* air sparging system are listed below.

- Air distribution (zone of influence)
- Depth of air injection
- Air injection pressure and flow rate
- Injection mode (pulsing)
- Injection wells
- Contaminant type and distribution

Air Distribution (Zone of Influence)

During the design of air sparging systems, it may be very difficult to define a radius of influence as it is used in pump-and-treat and/or soil venting systems. Due to the asymmetric nature of the air channel distribution and the variability in the density of channels, it is safer to assume a "zone of influence" than a radius of influence.

It becomes necessary to estimate the "zone of influence" of an air sparging point, similar to any other subsurface remediation technique, to design a full-scale air sparging system consisting of multiple points. This estimation becomes an important parameter for the design engineer to determine the number of required sparge points. The zone of influence should be limited to describing an approximate indication of the average of the furthest distance traveled by air channels from the sparge point in the radial directions, under controlled conditions.

The "zone of influence" of an air sparging point is assumed to be a cone; however, it should be noted that this assumption implies homogeneous soils of moderate to high permeability, which is rarely observed in the field. During a numerical simulation study on air sparging (Lundegard and Andersen, 1993), three phases of behavior were predicted following initiation of air injection. These are (1) an expansion phase in which the vertical and lateral limits of air flow grow in a transient manner, (2) a second transient period of reduction in the lateral limits (collapse phase), and (3) a steady-state phase, during which the system remains static as long as injection parameters do not change. The zone of influence of air sparging was found to reach a roughly conical shape during the steady-state phase.

Based on the inverted cone air-flow distribution model, many air sparging system designs are performed based on the "zone of influence" measured by conducting a field design test. Many applications require multiple zones to cover an entire area. When a hot spot or source area is under consideration for cleanup, it is prudent to design the air sparging system in a grid fashion (Figure 4). The grid should be designed with overlapping zones of influence providing complete coverage of the area under consideration for remediation. If an air sparging curtain is designed to contain the migration of dissolved contaminants, the curtain should be designed with overlapping zones of influence in a direction perpendicular to the direction of groundwater flow (Figure 4).

A properly designed test can provide valuable information. The limitations of time and money often restrict field evaluations to short duration single-well tests. Potential measuring techniques (Figure 5) of the zone of influence have evolved with this technology during the last few years.

1. *Measurement of the lateral extent of groundwater mounding* (Brown, 1992b; Brown et al., 1991): This was the earliest technique used during the very early days of implementation of this technology. It did not take long to conclude that the lateral extent of the mound is only a reflection of the amount of water displaced and does not correspond to the zone of air distribution.

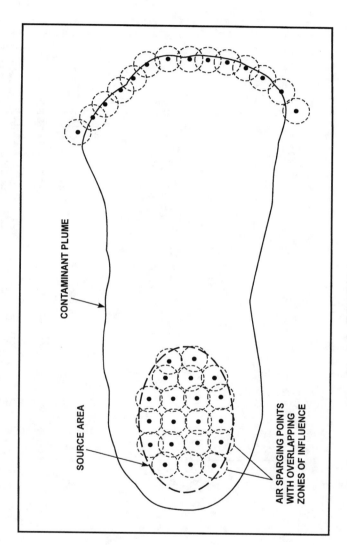

Figure 4 Air sparging points location in a source area and in a curtain configuration.

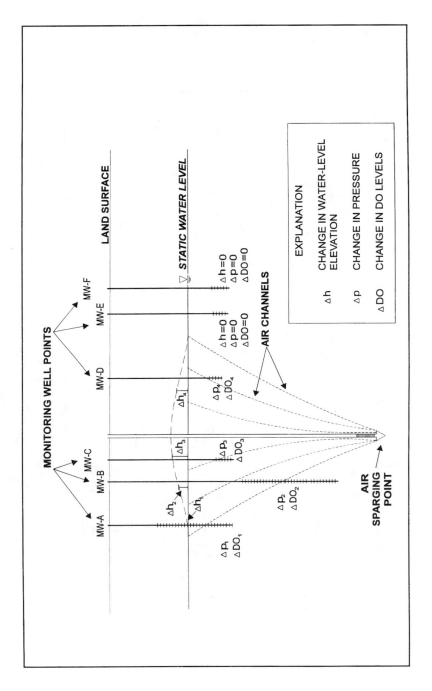

Figure 5 Air sparging test measurements.

2. *Increase in DO levels and redox potentials in comparison to pre-sparging conditions* (Brown, 1992b; Sellers and Schreiber, 1992; Marley et al., 1992): This concept lost its value when it was realized that the injected air travels in the form of channels more than as bubbles. In most cases the increased DO levels observed were due to the air channels directly entering the monitoring wells and not due to the diffusional transport of dissolved oxygen.

3. *Increase in head space pressure within sealed saturated zone monitoring probes which are perforated below the water table only:* This technique is the most widely used technique currently and is considered to be the most reliable in terms of detecting the presence of air pathways at a specific distance in the saturated zone.

4. *The use and detection of insoluble tracer gases, such as helium and sulfur hexafluoride* (Sellers and Schreiber, 1992): This technique is used similarly to the above method. The potential to balance the mass of the injected tracer and the amount of recovered tracer raises the level of confidence in the estimation of the total extent of channel distribution. The use of sulfur hexafluoride as a tracer has an advantage due to its solubility being similar to that of oxygen.

5. *Measurement of the electrical resistivity changes in the target zone of influence as a result of the changes in water saturation due to the injection of air:* This may be the most reliable method but the cost of this technique precludes it from being used widely.

6. *The actual reduction in contaminant levels due to sparging:* This evaluation gives an indication of the extent of the zone of influence in terms of contaminant mass removal, but the test has to be run long enough to collect reliable data.

Since cost and budgetary limitations influence how a field design test is performed, availability of resources will determine the type of method that is used. The most reliable method is the one which measures the changes in electrical resistivity due to changes in air/water saturation. The most cost-effective method is the one that determines the head space pressure within the saturated zone probes.

Depth of Air Injection

Among all the design parameters, depth of air injection may be the easiest to determine since the choice is very much influenced by the contaminant distribution. It is prudent to choose the depth of injection at least a foot or two deeper than the "deepest known point" of contamination. However, in reality, the depth determination is very much influenced by soil structuring and extent of layering, since injection below any impermeable or very permeable zones should be avoided. The current experience in the industry is mostly based on depths less than 30 ft (EPA Report 68-03-3409, 1992), below the water table. In theory, only the pressure available from the blower would limit the depth of application. The VES associated with the sparging system must also be considered when selecting the depth.

Air Injection Pressure and Flow Rate

The injected air will penetrate the aquifer only when the air pressure exceeds the sum of the hydrostatic pressure of the water column and the threshold capillary pressure, or the "air entry pressure". The air entry pressure is equal to the minimum capillary entry resistance for the air to flow into the porous medium. Capillary entry resistance is inversely proportional to the average diameter of the grains and porosity (Bohler et al., 1990; Lundegard and Andersen, 1993). Thus, the "air entry pressure" will be higher for fine-grained media and lower for coarse-grained media.

The injection pressure necessary to initiate *in situ* air sparging should be able to overcome the following:

1. The hydrostatic pressure of the overlying water column at the point of injection
2. The capillary entry resistance to displace the pore water, which depends on the type of sediments in the subsurface

The capillary pressure can be described (Ardito and Billings, 1990) under idealized conditions by the following equation:

$$Pc = 2 \ s/r$$

where

Pc = capillary pressure
s = the surface tension between air and water
r = the mean radius of curvature of the interface between fluids

This equation reveals that as r decreases the capillary pressure increases. Generally, r will decrease as grain size decreases; hence, the required pressure to overcome capillary resistance increases with decreasing sediment size.

Hence, the pressure of injection (P_i) in ft of water could be defined as:

$$P_i = H_i + P_a + P_d$$

where

H_i = saturated zone thickness above the sparge point (feet of water)
P_a = air entry pressure of formation (feet of water)
P_d = air entry pressure for the well, if a diffuser is used.

The air entry pressure is heavily dependent on the type of geology. In reality, the air entry pressure will be higher for finer sediments (1 to 10 ftH$_2$O) than coarser sediments (1 to 10 inH$_2$O).

The notion that higher pressures and flow rates correspond to better air sparging performance is not true. Increasing the injection rate to achieve a greater flow and wider zone of influence must be implemented with caution (Johnson et al., 1993; Lundegard and Andersen, 1993). This is especially true, during the startup phase due to the low relative permeability to air because of the low initial air saturation. The danger of pneumatically fracturing the formation under excessive pressures should also be taken into consideration in determining injection pressures. Hence, it is very important to gradually increase the pressure during system start-up.

The typical values of injected air flow rates reported in the literature (EPA Report 68-03-3409, 1992; Johnson et al., 1993) ranges from 1 cfm to 10 cfm. Injection air flow determinations are influenced more by the ability to recover the stripped contaminant vapors through a vapor extraction system by containing the injected air within a controlled air-distribution zone.

Injection Mode (Pulsing)

Direct and speculative information available in the literature indicates that the presence of air channels impedes, but does not stop, the flow of water across the zone of influence of sparging. The natural groundwater flow through a sparged zone of an aquifer will be slowed and diverted by the air channels due to changes in water saturation and thus relative hydraulic permeability. This potentially negative factor could be overcome by pulsing the air injection and thus minimizing the decrease in relative permeability due to changes in water saturation.

An additional benefit of pulsing likely will be due to the agitation and mixing of water as air channels form and collapse during each pulse cycle. This should also help to increase the rate of transport of contaminants in the bulk water phase towards the air channels, due to the cyclical displacement of water during pulsed air injection.

Injection Wells

Injection wells have to be designed in such a way to accomplish the desired distribution of air flow in the formation. Conventional design of an air sparging well under shallow "sparge depth" conditions (less than 20 ft) and deeper sparge depth conditions (greater than 20 ft) are shown in Figures 6 and 7, respectively. Schedule 40 or 80 PVC (polyvinyl chloride) piping and screens in various diameters can be used for the well construction. In both configurations, the sparge point can be installed by drilling a well to ensure an adequate seal to prevent short circuiting of the injected air. Hence, at large sites, the cost of installing multiple sparge points may prohibit the consideration of air sparging as a potential technology.

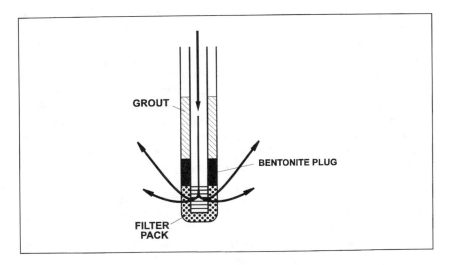

Figure 6 Schematic showing conventional design of an air sparging point for shallower applications.

Installation of air sparging points, with driven well points made out of small diameter (3/4- to 1 1/2-in.) 8- to 10-ft cast iron, flush-jointed sections (Figure 8) will help in making this technology more cost effective. However, in the absence of a sand pack, ground around the sparge points may clog up over a long period. Continuous expansion and collapse of the soils around the sparge point during the pulse cycles will act like a "sieving" action, thus finer sediments may accumulate around the sparge points and eventually clog them.

Contaminant Type and Distribution

Volatile and strippable compounds will be more amenable to air sparging. Nonvolatile but biodegradable compounds also will be addressed by this technique. This means that most petroleum hydrocarbons and chlorinated solvents can be treated with air sparging. Even compounds such as acetone and other ketones that cannot be treated with an air stripper above ground can be affected by air sparging due to the biological activity. A compound such as 1,4-dioxane, which is very soluble and nondegradable, will not be remediated.

There is no limit on the dissolved concentrations of contaminants in applying air sparging. For air sparging to be effective, the air saturation percentage and the radius and density of air channels are important factors for mass transfer efficiencies of both contaminants and oxygen. The rates of stripping and biodegradation are both limited by diffusion through water. It is not possible to optimize them separately.

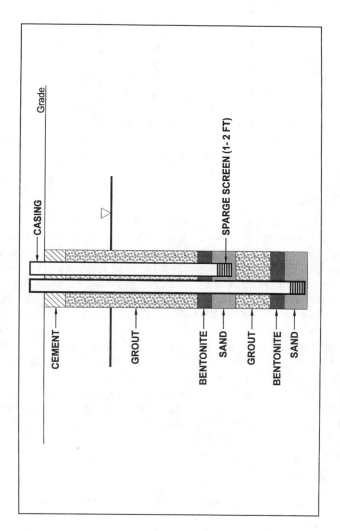

Figure 7 Diagram of a nested sparge well for deeper applications.

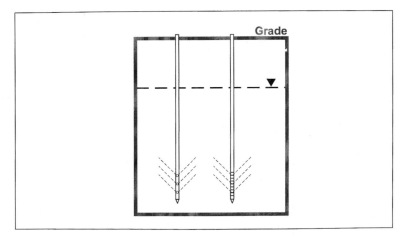

Figure 8   Small diameter air sparging well configuration.

PILOT TESTING

Because state-of-the-practice designing of *in situ* air sparging systems has not progressed beyond the "empirical stage", a pilot study should be considered to prove the effectiveness of the system as well as to gather the data necessary for full-scale design. The pilot study could be more appropriately defined as a field design study, since the primary objective would be to obtain site-specific design information. However, due to the still unknown nature of the mechanics of the process, the data collected from a pilot test should be treated with caution. The collected data should be valued as a means of overcoming any prior concerns regarding implementation of this technology. Since vapor extraction technology is complementary to *in situ* air sparging, pilot testing of the integrated system at the same time is highly recommended.

It is very important to perform a preliminary evaluation of the geologic and hydrogeologic conditions for the applicability of *in situ* air sparging prior to the pilot study. In addition, a thorough examination of the degree and extent of contamination should be performed.

The following evaluation prior to designing a pilot test will enhance the quality of data that could be collected.

The equipment set up used for an air sparging pilot test is similar to the one shown in Figure 1. The data that should be collected during the pilot study include the following engineering parameters.

Zone of air distribution: For any subsurface remediation system, this is a key design parameter since this would determine the required number of injection points. The zone of influence under various pressure and flow combinations should be measured. The methods to infer the zone of influence have been described in the Air Distribution section of this chapter and Figure 4.

Condition	Impact
Saturated zone soil permeability	Applicability — flow rate vs. pressure
Geologic stratification	Applicability — air distribution and flow pattern
Depth of contamination below the water table	Sparging depth (injection pressure)
Type of contaminant	Applicability — volatility/strippability/biodegradability
Extent of contamination	Applicability and efficiency (multiple sparge points)
Soil conditions above the water table	Ability to capture the stripped contamination by vapor extraction

Injection air pressure: This parameter is very much influenced by the depth of injection and subsurface geology. The required baseline pressure during the test should be equal to or just above the value necessary to overcome the sparging depth. The impact of any additional required pressure should be evaluated carefully in incremental steps, because excessive pressures may fracture the soils around the point of injection.

Injection flow rate: Evaluation of the injection flow rate should be governed by the ability to capture the stripped contaminant vapors and the net pressure gradient in the vadose zone. At a minimum, the air flow rate should be sufficient to promote significant volatilization rates and maintain dissolved oxygen levels greater than 2 mg/L. Typical injection flow rates are in the range of 1 to 10 cfm per injection point.

Mass removal efficiency: Another key objective during the field study should be to demonstrate the mass removal efficiency of the *in situ* air sparging process. This can be determined by measuring the net increase in contaminant levels in the vapor extraction system after the initiation of the air sparging system. Hence, the field test should be done as a sequential test in two phases.

The first phase should be to perform the vapor extraction test and monitor the effluent air levels under steady-state conditions. The air sparging is initiated during the second phase with continued monitoring of the contaminant levels in the vapor extraction system air stream. An increase in the contaminant level and the duration of increase would indicate the mass removal efficiency due to air sparging, Figure 9. The first phase of the test should be conducted to evacuate a minimum of 1.5 to 2 pore volumes of soil gas. The second phase of the test should be continued until we start to see a decline in concentrations in the effluent air stream or until 2 to 3 pore volumes of air have been injected into the affected saturated zone. NAPLs and geologic heterogeneities may prevent the concentrations during the second phase from declining. Extended pilot plants may be required if it is important to determine the total time required for full-scale sparging operations.

Determination of the increase in contaminant levels, due to air sparging, is important to evaluate the safety considerations of implementing this technology.

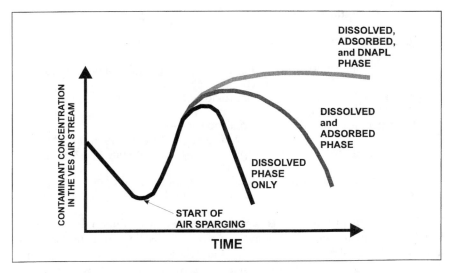

Figure 9   Contaminant removal efficiencies during a pilot test.

Continuous removal of the contaminants transferred into the vadose zone is very important. Buildup of these contaminants to explosive levels should be avoided at any cost. Hence, the air injection rate should be controlled in order to maintain a net negative pressure within the target area.

LIMITATIONS

At first glance, *in situ* air sparging appears to be a simple process: injection of air into a contaminated aquifer below the water table with the intent of volatilizing VOCs and providing oxygen to enhance biodegradation. It is also important to know when not to apply this technology. The following provides a brief summary of the conditions under which conventional application of this technology is not recommended:

- Tight geologic conditions with hydraulic conductivities less than 10^{-3} cm/sec. The vertical passage of the air may be hampered, and the potential for the lateral movement will be increased, as will be the potential for inefficient removal of contaminants. Conventional forms of air sparging should be evaluated with extreme caution under these conditions.
- Heterogeneous geologic conditions, with the presence of low permeability layers overlying zones with higher permeabilities. The potential for the enlargement of the plume exists again due to the inability of the injected air reaching the soil gas above the water table.
- Contaminants present are nonstrippable and nonbiodegradable.

- Mobile free product has not been removed or completely controlled. Air injection may enhance the uncontrolled movement of this liquid away from the air injection area.
- Air sparging system cannot be integrated with a vapor extraction system to capture all the stripped contaminants. Sometimes the stripped contaminants can be biodegraded in the vadose zone if optimum conditions are available. It is very difficult to design for these conditions and this design method is rarely accepted.
- The structural stability of nearby foundations and buildings may be in jeopardy.
- Nearby basements, buildings, or other conduits may be affected by the vapors created from the sparging.

MODIFICATIONS TO CONVENTIONAL AIR SPARGING APPLICATION

For the purposes of discussion in this book, conventional application of air sparging is illustrated in Figure 1. Due to the geologic and hydrogeologic conditions encountered at many sites across the country, this form of application may have to be limited to only 25% of the remediation sites (Geraghty & Miller). However, the concept of using air as a carrier for removing contaminant mass from the saturated zone still remains very attractive and cost effective compared to currently available options. There are several modifications to the conventional air sparging design that can be used to extend its application.

Horizontal Trench Sparging

This technique was developed to apply air sparging under less permeable geologic conditions. When the hydraulic conductivities (in the horizontal direction) are less than 10^{-3} cm/sec, it is prudent to be cautious and not inject any air directly into the water saturated formations.

The primary focus in this modified approach is to create an artificially permeable environment in the trenches placed across the direction of contaminant migration. As the contaminated water travels through a trench, the strippable VOCs will be removed from the ground water and captured by the vapor extraction pipe placed above the water table. The configuration of a horizontal trench sparging system is shown in Figure 10.

Due to the extremely slow groundwater velocities under conditions where this technique may be preferred, the residence time of the moving ground water in the trench will be high. Hence, the injection of the air does not have to be continuous and a pulsed mode of injection can be instituted. When biodegradable contaminants are present, the trench can be designed to act like an *in situ* fixed-film bioreactor and thus the rate of injection of air can be decreased. The need for effluent air treatment can be avoided under these circumstances. Injection of nutrients, such as nitrogen and phosphorus, may enhance the rate of biodegradation in the trench.

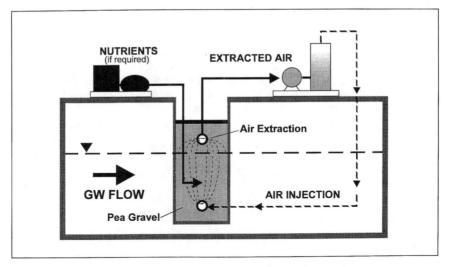

Figure 10   Horizontal trench sparging (section view).

The treated ground water leaving the trench will be saturated with dissolved oxygen and nutrients and hence will enhance the degradation of dissolved and residual contaminants downgradient of the trench also. If the primary focus of remediation is containment only, this concept can be implemented as a passive contaminant technique as shown in Figure 10 or, depending on the need to clean up the site faster, it can be implemented as shown in Figure 11. The biggest limitation of this technique will be the total depth of the trench. Total depths beyond 25 ft may preclude the implementation of this technique due to the need to deal with a large volume of contaminated soils.

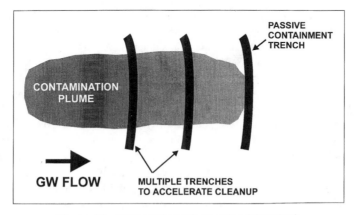

Figure 11   Horizontal trench sparging (plan view).

Since the depths will be limited to less than 25 ft, air injection into the trench can be accomplished with a blower instead of a compressor. Hence, the extracted air can be treated with a vapor treatment unit (probably vapor phase granular activated carbon (GAC) due to the low levels of mass expected) and re-injected back into the trench as shown in Figure 10. This configuration will eliminate the need to take continuous air samples.

In Well Air Sparging

This modification was developed as a means to use air as the carrier of contaminants and also, at the same time, to overcome the difficulties of injecting air into "non-optimum" geologic formations. This configuration, shown in Figure 12, also overcomes the difficulties of installing the trenches described in the previous section. Proprietary techniques that use this basic concept are also covered in Chapter 10.

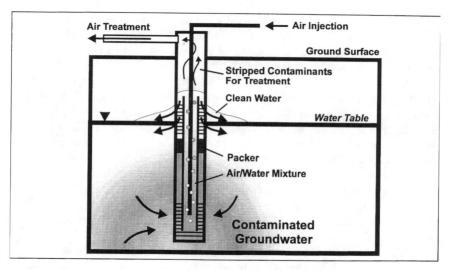

Figure 12   In-well air sparging (patent pending).

The injection of air into the inner casing, Figure 12, induces an "air lifting effect", which is limited only to the inner casing. Hence, the water column inside the inner casing will be lifted upwards (in other words, water present inside the inner casing will be pumped) and will flow over the top of the inner casing shown in Figure 12. As a result, contaminated water will be drawn into the lower screen from the surrounding formation and also will be "air lifted" in the inner tube continuously.

Due to the mixing of air and contaminated water, as the air/water mixture rises inside the inner tube, strippable VOCs will be air stripped and captured for treatment (Figure 10). Treated, clean water that spills over the top of the inner casing will be released back into the formation via the top outer screen. This approach completely eliminates the need for extracting the water for above-ground treatment.

The re-injected water, saturated with dissolved oxygen, will enhance the biodegradation of any biodegradable contaminants present in the saturated zone. The need to inject nutrients inside the well can be evaluated on a site-specific basis.

Biosparging

As discussed in previous sections, injection of air into water-saturated formations has a significant benefit in terms of delivery of oxygen to the microorganisms for *in situ* bioremediation. If delivery of oxygen for the biota is the primary objective for air injection, the volume of air flow does not have to be at the same level required to achieve stripping and volatilization. Control of air channel formation and distribution and capturing of the stripped contaminants also become insignificant under these circumstances. Application of this technique to remediate a dissolved plume of acetone, which is a nonstrippable but extremely biodegradable compound, will be appropriate.

Injection of air at very low flow rates (1/4 to less than 2 cfm per injection point) into water-saturated formations to enhance biodegradation is termed "biosparging". Limitations caused by geological formations also become insignificant since the path of air channels can be allowed to follow the path of least resistance. However, it has to be noted that the time required to increase the dissolved oxygen levels in the bulk water depends on the time required for the diffusion of O_2 from the air channels into the water surrounding the channels. Hence, caution has to be exercised in terms of evaluating the changes in dissolved oxygen (DO) levels after the initiation of biosparging. It is not uncommon to assume that the observed increase in DO levels in monitoring wells is due to changes in the bulk water. Direct introduction of air into the monitoring wells due to an air channel being intercepted could also be a reason for increased DO levels in monitoring wells. Chapter 3, *In Situ* Bioremediation, provides more details of the designs, operations, and associated costs for biosparging.

CLEANUP RATES

To date, there are no reliable methods for estimating groundwater cleanup rates. A mass removal model for *in situ* air sparging using air stripping as the only mass transfer mechanism has been reported (Marley et al., 1992). But, this model was based on the premise that injected air travels in the form of bubbles and, hence, the reliability of this model may be questionable.

It has been established that the primary mode of travel of injected air at the majority of the sites is in the form of air channels. In the presence of air channels, the rate of mass transfer will be limited either by kinetics of the mass transfer at the interface or by the rate of transport of the contaminant through the bulk water phase to the air/water interface. Based on these assumptions, reaching nondetect-able levels may be possible only with biodegradable contaminants (see Chapter 1 for a full discussion on the geologic limitations of air as a carrier).

Cleanup times of less than 12 months to 3 years have been achieved in many instances. Reports in the literature indicate that sites that have implemented air sparging have often met groundwater cleanup goals in less than 1 year (EPA Report 68-03-3409, 1992; Brown, 1992b; Marley et al., 1992; Kresge and Dacey, 1991; Ardito and Billings, 1990). However, it should be noted that, at most of these sites, the cleanup goal was around 1 mg/L for total BTEX and that BTEX compounds are very biodegradable. The required cleanup times for a site will depend on the following:

- Target cleanup levels
- Extent and phases of contamination
  - Contaminant mass present in the saturated zone and the capillary fringe
  - Extent of dissolved and sorbed phase contamination
  - The presence or absence of a dense non-aqueous phase liquid (DNAPL)
- Strippability, volatility, and biodegradability of contaminants present
- Solubility and partitioning of the contaminants
- Geologic conditions
  - Percentage of air saturation
  - Density of air channels
  - Size of the air channels

The air saturation and the size of the air-filled regions have the largest effect on the mass transfer rates. In other words, a lot of air-filled spaces that are finely distributed will promote a faster mass transfer. Numerical analyses show that the air saturation must be greater than 0.1%, and the size of air channels must be on the order of 0.001 m in order for sparging to be successful (Mohr, 1995). See Chapter 2 for a full discussion on the life cycle of remediations.

REFERENCES

Ahlfeld, D.P., Dahmani, A., and Wei, Ji. "A Conceptual Model of Field Behavior of Air Sparging and Its Implications for Application," *Groundwater Monitoring and Remediation,* vol. 24, no. 4, pp. 136–139, Fall 1994.

Angell, K.G. "In Situ Remediation Methods: Air Sparging," *The National Environmental Journal,* pp. 20–23, January/February, 1992.

Ardito, C.P. and Billings, J.F. "Alternative Remediation Strategies: The Subsurface Volatilization and Ventilation System." Proc. Petroleum Hydrocarbons and Organic Chemicals in Groundwater: Prevention, Detection, and Restoration Conf., API/NWWA, Houston, TX, 1990.

Bohler, J., Brauns, J., Hotzl, H., and Nahold, M. Air Injection and Soil Air Extraction as a Combined Method for Cleaning Contaminated Sites — Observations from Test Sites in Sediments and Solid Rocks, in *Contaminated Soil,* F. Arendt, M. Hinsevelt, and W.J. van der Brink, eds., Dordrecht, The Netherlands: Kluwer Academic Publishers, 1990.

Brown, R.A. *Air Sparging: A Primer for Application and Design,* Subsurface Restoration Conference, U.S. Environmental Agency, 1992a.

Brown, R.A. *Treatment of Petroleum Hydrocarbons in Groundwater by Air Sparging,* B. Wilson, J. Keeley, and J.K. Rumery, eds., Section 4, Research and Development, RSKERL — Ada, OK, U.S. Environmental Protection Agency, November 1992b.

Brown, R., Herman, C., and Henry E. "The Use of Aeration in Environmental Cleanups," presented at HAZTECH International Pittsburgh Waste Conference, Pittsburgh, PA, 1991.

EPA, "Evaluation of the State of Air Sparging Technology," Report 68-03-3409, U.S. Environmental Protection Agency, Risk Reduction Engineering Laboratory, Cincinnati, OH, 1992.

Geraghty & Miller, Inc., Air Sparging Projects Data Summary.

Howard, P.H., et al. *Handbook of Environmental Degradation Rates,* Chelsea, MI: Lewis Publishers, 1991.

Johnson, R.L., Center for Groundwater Research, Oregon Graduate Institute, Beaverton, Oregon, personal communication, April 1995.

Johnson, R.L., Johnson, P.C., McWharter, D.B., Hinchee, R.E., and Goodman, I. "An Overview of In-Situ Air Sparging," *Groundwater: Monitoring and Remediation,* vol. 23, no. 4, pp. 127–135, Fall 1993.

Kresge, M.W. and Dacey, M.F. "An Evaluation of In-Situ Groundwater Aeration," Proc. Ninth Annual Hazardous Waste Materials Management Conference/International Conf., Atlantic City, NJ, 1991.

Lundegard, P.D. and Andersen, G. "Numerical Simulation of Air Sparging Performance," Proc. Petroleum Hydrocarbons and Organic Chemicals in Groundwater: Prevention, Detection, and Restoration Conf., Houston, TX, 1993.

Lyman, W.J., Reehl, W.F., and Rosenblatt, D.H. *Handbook of Chemical Property Estimation Methods,* New York: McGraw-Hill, 1992.

Marley, M.C., Li, F., and Magee, S. "The Application of a 3-D Model in the Design of Air Sparging Systems," Proc. Petroleum Hydrocarbons and Organic Chemicals Groundwater: Prevention, Detection, and Restoration Conf., Houston, TX, 1992.

Marley, M.C., Walsh, M.T., and Nangeroni, P.E. "Case Study on the Application of Air Sparging as a Complementary Technology to Vapor Extraction at a Gasoline Spill Site in Rhode Island, Proc. Hazardous Materials Control Research Institute 11th Annual National Conf., Washington, D.C., 1990.

Mohr, D.H. Mass Transfer Concepts Applied to In Situ Air Sparging, Third International Symposium on In Situ and On-Site Bioreclamation, April 24-27, 1995, San Diego, CA, 1995.

Sellers, K. and Schreiber, R. "Air Sparging Model for Predicting Groundwater Cleanup Rate," Proc. Petroleum Hydrocarbons and Organic Chemicals in Groundwater; Prevention, Detection, and Restoration Conf., Houston, TX, 1992.

Wei, J., Dahmani, A., Ahlfeld, D.P., Lin, J.D., and Hill, E. "Laboratory Study of Air Sparging: Air Flow Visualization," Groundwater: Monitor. Remed., Fall, 1993.

7 AIR TREATMENT FOR *IN SITU* TECHNOLOGIES

Sami Fam

INTRODUCTION

We introduced in Chapter 1 the concept that most *in situ* treatment processes were simply a switch from water to air as the carrier. This chapter will look at treating the air carrier as it is brought above ground.

Above-ground vapor treatment of emissions from soil vapor extraction, air sparging, and air stripping applications often represents the largest portion of the overall cost of implementing these technologies. Figure 1 represents a pie chart of overall project costs associated with a vapor extraction system operating for 3 yrs at a vapor recovery rate of 300 cfm and a declining influent concentration from 2000 (hydrocarbon vapors) to 5 ppm over the project lifetime. It is assumed that vapors are treated using catalytic oxidation for the first 2 yrs, and that granular activated carbon (GAC) is used for treatment during the last year of treatment. Figure 1a shows that air emission control and operating and maintenance (O&M)-related costs may be over 50% of overall project costs. Figure 1b shows that O&M costs can be reduced significantly using remote sentry/telemetry capabilities for optimizing system performance.

Due to the magnitude of air emission control costs, the design engineer must carefully evaluate and select the most appropriate technology. Control technology selection must consider several criteria that will be introduced in this chapter:

- Regulatory requirements
- Overall mass of volatile organic components (VOCs) to be treated
- Anticipated decline rate of VOC concentrations over project lifetime (life cycle design)
- Citing considerations
- Utility availability

0-87371-995-6/96/$0.00+$.50
© 1996 by CRC Press, Inc.

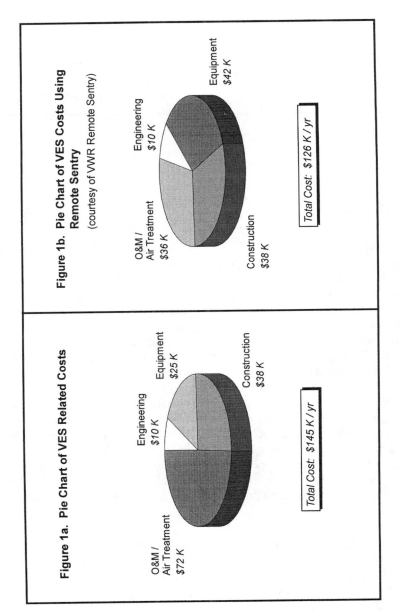

Figure 1   Pie chart of overall VES project costs.

- Organic and inorganic composition of process vapor stream
- Other project-specific considerations

The most common air emission control technologies can be classified as adsorptive, oxidative, or biological. The adsorption-based technologies include off-site regenerable/disposable vapor phase GAC, on-site steam-regenerable GAC, and on-site regenerable macroreticular resin systems. Oxidative technologies include catalytic oxidation and thermal oxidation. Biological-based systems have gained attention in the last few years and have become commercially available. Less commonly utilized technologies include scrubbing, vapor compression, ultraviolot/ozone oxidation, and refrigeration. The most commonly utilized technologies will be introduced in this chapter.

DESIGN CRITERIA

Regulatory Requirements

Vapor emissions from site remediation activities generally are not permanent sources of discharge. The short duration of the emission may exempt its permitting and control in some states. Often, however, in cities or states where overall air quality standards are not met (nonattainment areas) or in states with strict emission control standards, permitting and vapor treatment is required.

Emission requirements are quite variable within the different states. Emission requirements may be based on total mass emissions per hour or per day. Mass-based emission criteria may be for total VOCs or may be compound specific. The design engineer must select the air emission control system based upon the most limiting criteria. For example, VES emissions for a gasoline release contain a variety of compounds. If air emission standards require a maximum 15 lbs/day total VOC limit and an overall benzene limit of 1 lb/h, then system design must be based upon the limiting regulatory requirement. In this example, the limiting criteria is total VOC emissions, since benzene is usually 5% by mass of gasoline.

Alternatively, emission control criteria may simply require that the emission rates not cause an exceedance in ambient air quality or other risk-based criteria. This generally requires that the point source of discharge be modeled using air dispersion techniques (Bethea, 1978). Several states impose both mass emission and concentration criteria. Some states require that all emissions be treated using "best available control technology" (BACT) regardless of the magnitude of the emissions (this requirement calls for emission control even if the process stream already complies with mass emission requirements.) Therefore, the first thing that the design engineer must do is to acquire the local and state regulations before trying to design a vapor treatment system. Based upon life cycle design (Chapter 2), emission rates decline during site remediation. Permit preparation should account for this temporal change.

It is plausible to limit the site operation (hours per day or number of extraction wells) to stay within permitting limits (without treatment) until emission rates drop as the cleanup progresses. This approach will likely increase the site remediation time frame, and the design engineer must conduct a cost-benefit analysis in order to justify the merits of this phased start-up method.

Mass of Contaminants

An estimation of the total mass of VOCs that may be recovered by the remediation system is a requirement prior to determining the appropriate treatment technology. This is particularly true for adsorbent-based treatment systems. For example, if 1000 lbs of gasoline are known to be in the subsurface, and one expects that 65% of the mass will be recovered by vapor extraction, 30% will be biodegraded, and 5% will not be recovered (ratios are based upon empirical projection), an estimate can be formulated for expected adsorbent consumption. Assuming a 7% by weight adsorption capacity for GAC, then approximately 9300 lbs of GAC will be required. The cost of other technologies can also be estimated based upon the mass of VOCs and expected flow rates.

There are several simple methods to determine the mass of VOCs in the subsurface. Once this is known, an estimate can be made of the amount (percentage) that is expected to be extracted for above-ground treatment. Often the final estimate is based upon an average of the various estimation methods. An excellent starting point is direct knowledge of the amount of contaminants released. Time is also an important factor in that spills will weather and naturally degrade (see Chapter 3).

The total mass may then be estimated by using soil contaminant concentrations, groundwater concentrations, soil gas concentrations, and non-aqueous phase liquid (NAPL) thickness at the various locations across the site (see Equation 1 in Chapter 4). The use of weighted-average methods (concentration and expected flow from each zone) and subdivision of the site into small quadrants (based on the available data) will yield more accurate mass estimates.

It should be noted that soil analytical methods often underestimate the amount of adsorbed VOCs due to significant losses during the sampling procedures (EPA, 1991). Use of methanol extraction and preservation methods often can lead to soil contaminant levels that may be one to two orders of magnitude higher than conventional methods.

Finally, in instances where limited information is available, gross estimates of the total mass of contamination in the subsurface may be evaluated using partitioning coefficients. For example, if no soil contamination data is available, groundwater data, knowledge of the compound octanol/water partition coefficient and soil organic content can be used to estimate the amount of VOCs adsorbed to the soil (Equation 3 in Chapter 4). It should be stated that these equations assume equilibrium conditions persist in the subsurface. Non-equilibrium conditions generally dominate in the subsurface, and partitioning-based calculations underestimate the adsorbed mass. This is because a large portion of the mass may be restricted from being in equilibrium with the surrounding soil vapor/groundwater due to non-equilibrium type adsorptive or mass transfer limitations (Brusseau, et al., 1991).

Life Cycle Emission Concentration

Design of cleanup strategies to accommodate the life cycle of the project has been emphasized several times in this book. This is particularly true for the treatment of vapor emissions from a vapor extracting system (VES), whose concentrations may drop four orders of magnitude over a project lifetime (Figure 2). Emission control technology selection is more significantly affected by concentration and flow rate for vapor phase treatment than for liquid phase treatment. For example, an air stripper generally will be chosen for the treatment of 100 ppm or 100 ppb of ground water contaminated by BTEX compounds. This choice will be made for almost any flow rate. On the other hand, an emission stream of 10,000 ppm vapors from a VES stack (300 cfm), is best treated by a thermal oxidizer. As the concentrations drop to 1000 ppm, the vapors are best treated by catalytic oxidation. At influent concentrations of 20 ppm, the optimal choice may be GAC. The original design must encompass all of these criteria, not just the initial influent concentration.

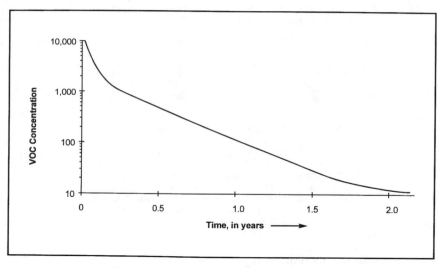

Figure 2   Life cycle curve for VES emissions.

This dynamic need to modify treatment technologies necessitates foresight from the design engineer for vapor emission system design. The systems must be designed and installed with sufficient flexibility to allow for future modifications. For example, at a site where VES emissions are expected to be above 300 ppm for 6 months and then drop off rapidly (typical of a small service station, limited-spill situation), a catalytic oxidizer may be rented for the first 6 months, and subsequently GAC may be installed at the site. The treatment system citing must therefore accommodate future GAC installation within the treatment building and a small concrete pad outside the structure for placement of the catalytic oxidation unit.

Typically, the engineer needs to predict the decline curve for the emissions from the air treatment system and subsequently prepare a cost analysis spreadsheet for the various options at varying concentrations. A typical cost analysis table is shown in Table 1. Modeling of the remedial system performance to predict the decline curve may be conducted. In many instances, this modeling is not performed, and empirical methods (fitting the concentration decay to an empirical logarithmic decay equation over a time period based upon past experience) are used for prediction of the decline curve. The use of empirical methods is generally acceptable in the consulting industry for purposes of air emission selection due to the costs of modeling and its inherent uncertainties. For example, it is not critical to know whether a catalytic oxidizer will run for 6 or 7 months before switching to GAC; what is important is the ability to plan for and switch to GAC.

Table 1   Cost Analysis Spreadsheet for Vapor Treatment Costs

Influent concentration	VOC per day	GAC cost/day ($)	Catalytic oxidation cost/day ($)
50 ppm	1.47	59	136
100 ppm	2.93	117	128
200 ppm	5.86	234	120
500 ppm	14.65	586	112
1,000 ppm	29.31	1,170	110
2,000 ppm	58.62	2,344	100
3,000 ppm	87.93	3,516	100

Note: Cost assumptions — 1 ppm = 3.26 mg/m³ (Benzene). 100 cfm operation for 24 hours per day. GAC adsorption capacity = 15% by weight. Carbon costs = $6/lb (new plus regeneration and changeout cost). Catalytic oxidation unit rental is $3000/month. Catalytic oxidation power consumption is $350/month (at 500-ppm influent); assume costs are slightly higher at lower concentrations, slightly lower at high concentrations.

Citing and Utility Considerations

There are several citing considerations that need to be evaluated prior to treatment technology selection. Some of these constraints are enumerated below:

- Availability of utilities
- Utility cost analysis
- Access issues relating to O&M
- Aesthetic issues
- Proximity to homes and buildings
- Winterization
- Other site specific considerations

The availability of utilities and their ability to accommodate the treatment equipment must be carefully evaluated. Most utiity evaluations are for thermal and catalytic oxidation. For example, if natural gas is to be selected, it must be available in sufficient pressure to be utilized by the treatment equipment. In old residential neighborhoods, natural gas lines may not have sufficient pressure for adequate operation of some thermal oxidation units. Electrical power must be available in the appropriate phase and voltage to power the equipment. At remote sites, where utility availability is limited, propane tanks can be utilized. The design engineer needs to conduct a cost analysis in order to choose the most appropriate power source (natural gas, electrical, propane, oil, etc.) for powering the treatment unit. When available, natural gas tends to be the lowest cost option in many locations. The use of propane, in addition to increased cost per BTU (generally 1.5 to 2 times higher than natural gas), also presents other operational problems such as increased fouling of burner components, as well as logistical problems created when scheduling fuel deliveries.

Remedial systems are unplanned installations. Sites and neighborhoods are obviously developed without planning for a potential remedial system installation. This unplanned remedial system, therefore, needs to be located to accommodate several factors that may sometimes be conflicting. It must be located to attain permitting, meet regulatory stack height and air dispersion requirements, fit in with the natural setting, and be accessible for routine operation and maintenance. Concurrently, the system must not be offensive to neighbors, and it must have a stack height that meets local zoning laws.

TREATMENT TECHNOLOGIES

Adsorption-Based Treatment Technologies

Adsorption is a process by which material accumulates on the interface between two phases. In the case of vapor phase adsorption, the accumulation occurs at the air/solid interface. The adsorbing phase is called the "adsorbent" and the substance being adsorbed is termed an "adsorbate". It is useful to distinguish between physical adsorption, which involves only relatively weak intramolecular bonds, and chemisorption, which involves essentially the formation of a chemical bond between the sorbate molecule and the surface of the adsorbent. Physical adsorption requires less heat of activation than chemisorption and tends to be more reversible (easier regeneration).

Granular activated carbon is the most popular vapor phase adsorbent in the site remediation industry. A number of new synthetic resins, however, have shown increased reversibility and have higher adsorption capacities for certain compounds.

The most efficient arrangement for conducting adsorption operations is the columnar, continuous-plug flow configuration known as a "fixed bed". In this mode, the reactor consists of a packed bed of adsorbent through which the stream under treatment is passed. As the air stream travels through the bed, adsorption takes place and the effluent is purified (Figure 3).

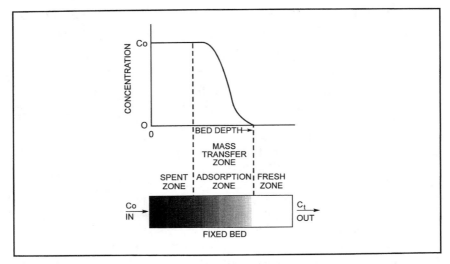

Figure 3   Concentration profile along an adsorbent column. (From Noll, K.E., Vassilios, G., and Hou, W.S., Adsorption *Technology for Air and Water Pollution Control,* Chelsea, MI: Lewis Publishing, 1992. With permission.)

The part of the adsorption bed that displays the gradient of adsorbable concentration is termed the mass transfer zone (MTZ). The amount of adsorbate within the bed changes with time as more mass is introduced to the adsorbent bed. As the saturated (spent or used) zone of the bed increases, the MTZ travels downward and eventually exits the bed. This gives rise to the typical effluent concentration vs. time profile, called the "breakthrough curve" (Figure 4). The reader is referenced to several textbooks for adsorption theory, multicomponet effects, isotherm description, and modeling (Noll et al., 1992, Faust and Aly, 1987). This basic knowledge of adsorption theory is critical to proper understanding and selection of the various adsorbents.

Off-Site Regenerable/Disposable Gas Phase Activated Carbon

Gas phase granular activated carbon is an excellent adsorbent for many VOCs commonly encountered in vapor extraction, air sparging, vacuum-enhanced recovery, and conventional groundwater extraction. The adsorption capacity of GAC is often quantified as the mass of contaminant that is adsorbed per pound of GAC. This nominal adsorption capacity is a useful guide for pure compound adsorption but can be misleading when complex mixtures of VOCs are treated. Breakthrough, or GAC bed life, is defined when breakthrough occurs for the compound most difficult to adsorb. In instances where multicomponent mixtures are present, the adsorption capacities for each compound are generally lower than for pure compounds. Isotherm data (milligrams of adsorbate removed per gram adsorbent at a constant temperature) and other product-specific data are generally available for contaminants of interest from carbon vendors as well as in the

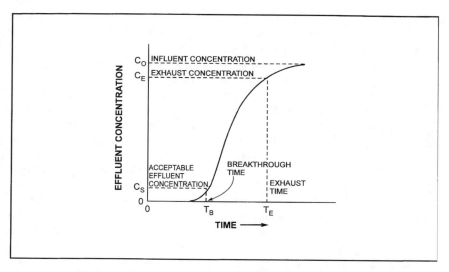

Figure 4   Breakthrough curve for a typical adsorber column.

published literature. Pilot testing for GAC feasibility is rarely conducted, except in instances where complex mixtures of VOCs are encountered.

Granular activated charcoal, a good adsorbent for hydrocarbon origin VOCs and some chlorinated VOCs, has limited adsorption capacity for ketones and generally poor adsorption of volatile alcohols. Table 2 shows typical adsorption removal efficiencies for a variety of VOCs by GAC under constant temperature and moisture conditions (as stated in the table).

Table 2   Adsorption Capacity of GAC for Some Common VOCs

	Carbon capacity[a] (at 10 ppmv)	Pounds organic per 100 lbs GAC (at 100 ppmv)
Benzene	13	19
Carbon Tetrachloride	20	33
Methylene chloride	1.3	2.7
Toluene	21	27
Trichloroethlene	19	33

[a] Carbon capacities are based on Calgon BPL carbon at 70°F and 1 atm. Values adjusted to reflect usage per 100 lbs GAC.

Note: ppmv = part per million by volume.

Source: Calgon Bulletin #23-77a, Pittsburgh, PA, February 1986.

Granulated activated charcoal adsorption capacity is significantly enhanced if the relative humidity of the vapor stream is kept low. The use of water knockouts, demisting, desiccants and air stream temperature adjustments are therefore common pretreatment steps to enhance GAC performance. Adsorption capacity may be as much as ten times higher for a low humidity stream than for a humid air stream. This is particularly true for lower concentrations of VOCs. The humidity effects are less pronounced at higher VOC concentrations (Figure 5). Temperature elevation after water demisting/desiccation increases the amount of moisture that the air stream can sustain, thus reducing the relative humidity. It should be noted that the adsorption vessels must also be kept warm so as to avoid water condensation on the GAC and for the air stream to maintain the elevated relative humidity.

The use of off-site regenerable/disposable GAC for emissions control is often limited to instances where the mass loading is low and, therefore, GAC consumption is low. As a general guideline, adsorption capacities for adsorbable VOCs are in the range of 2 to 20% by weight. Costs of GAC are in the range of $3/lb (if purchased in canisters; $1/lb if purchased as carbon only) plus an additional $0.5 to 1.0/lb for regeneration/disposal. Vessels of GAC may be purchased as canisters (200-lb size; the vessel and GAC are replaced after consumption), larger replaceable plastic/fiberglass canisters, or as conventional steel vessels wherein only the GAC is replaced. Vessels are most commonly used in series in order to allow for effluent stream sampling between vessels to more accurately predict breakthrough times (Figure 6). Single vessels may also be used in conjunction with vapor effluent detectors that can shut down the system or switch spent vessels to standby unused vessels. GAC vessel/blower selection must account for head losses through the carbon system to ensure maintenance of the desired air flow.

On-Site Regenerable GAC

Liquid phase GAC is difficult to regenerate at low temperatures due to adsorption of background metals and total organic carbon (TOC; typically naturally occurring humic and fulvic acids). In the vapor phase, the GAC does not have to deal with the metals nor the nonvolatile TOC and is therefore amenable to on-site low temperature regeneration. Air emission control using steam regenerable GAC generally utilizes a two-bed system whereby one bed is being utilized, while the second bed is either being regenerated or is in a stand-by mode. Regeneration is accomplished by passing low pressure steam through the carbon vessel, which desorbs the contaminates (due to the high temperature). The contaminated steam is subsequently cooled and separated from the non-aqueous organics by gravity. The condensed organics require disposal, whereas the contaminated steam may undergo water treatment (especially if a groundwater treatment system exists on-site) or may also require off-site disposal. If the condensed steam is treated on-site, it should be metered into the groundwater treatment system, since it is generally much more contaminated than the ground water.

If regenerable systems are used for adsorption of chlorinated VOCs, the vessels should be lined or made of an acid resistant material. The adsorption/regeneration cycle results in formation of some breakdown hydrochloric (HCl) acid within the vessels. The HCl formation will reduce the pH of the condensed steam.

Regenerable GAC units are also available by nitrogen regeneration. This is particularly useful in minimizing steam disposal and eliminating problems with HCl formation during steam regeneration. Regenerable beds usually will have a performance lifetime, since adsorption capacity tends to diminish with continued regeneration (typically decay to 70% of original capacity). System lifetime ranges

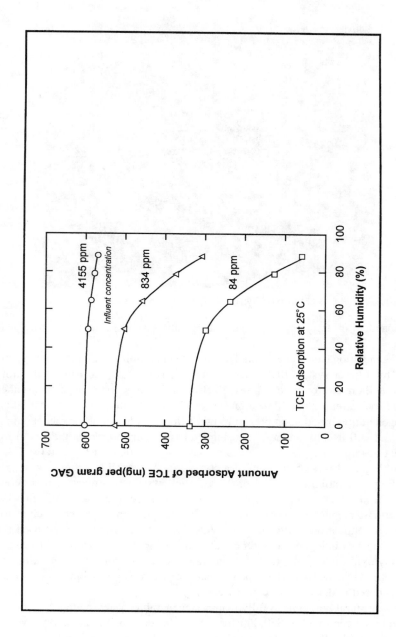

Figure 5 Effects of relative humidity on TCE adsorption by GAC.

Figure 6   Typical GAC system in series.

are dependent on frequency of regeneration but are typically in the 3- to 7-year range.

Regeneration can (1) be manual during site operation and maintenance visits, (2) be by a timer system that starts up the boiler/regeneration prior to the time of expected breakthrough, or (3) use an effluent detector system (typically either a flame ionization or photoioization detector) which initiates vessel alternation and regeneration based upon breakthrough. After completion of the regeneration, the vessel will usually undergo a drying cycle in order to prepare the vessel for the next adsorption cycle. Figure 7 provides a schematic of a regenerable GAC system. Figure 8 is a photograph of a commercially available system.

The regeneration capability of the GAC allows for treatment of more contaminated air streams than is possible with off-site regenerable GAC. In VES applications, regenerable GAC systems are often thought of as best suited for high flow applications with moderate VOC loadings, and they can be used for the treatment of halogenated and nonhalogenated air streams. During adsorption of highly oxidizable VOCs (ketones) in the presence of ozone or other oxidizing agents, the GAC bed may be prone to bed fires. Fire-suppression systems may be considered under these circumstances.

As a general guideline, a fully manual regenerable system with a boiler may be purchased for roughly $50,000 for a 500 cfm dual bed application. System automation with effluent detector will generally add approximately $40,000.

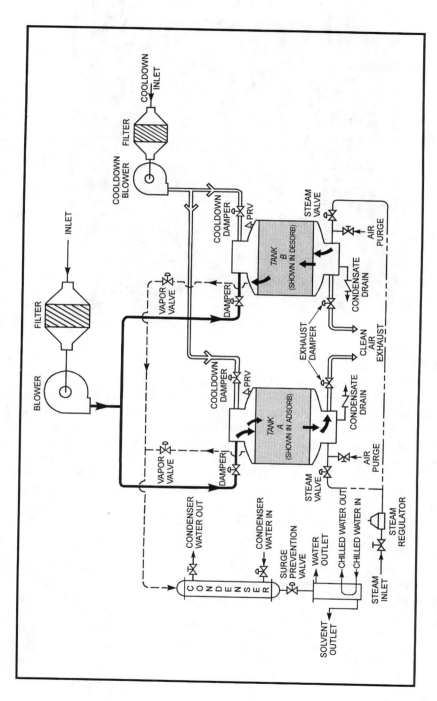

Figure 7　Regenerable GAC system schematic. (Adapted from Westport Environmental Systems, Westport, MA.)

Figure 8   Regenerable GAC system.

Resin Adsorption Systems

The use of adsorbent resins for water treatment began in the 1960s with the introduction by Rohm and Hass of their macroreticular absorbent adsorbent (Faust and Aly, 1987). These synthetic resins were designed to have low adsorption for background TOC and metals and thereby enable the liquid phase adsorbent to better adsorb the VOCs. Elimination of TOC and metal adsorption would also reduce biofouling (less TOC for microbial growth). The liquid phase resins were also designed to be steam regenerable since metals and TOC fouling would not impact regeneration. For a variety of economic and performance reasons, liquid phase adsorbent resins have gained limited acceptance; however, new resins by both Rohm and Hass (Philadelphia, PA) and Dow Chemical (Midland, MI) within the past 3 years may make liquid phase resin adsorption more economically competitive.

It had been commonly prescribed that these resins be pre-wet prior to their use. In the 1980s, it was observed that the resins have an almost similar adsorption capacity in their pre-wet condition (Yao and Tien, 1990; Rixey and King, 1989). Most recently, in the early 1990s, a vendor (Purus Inc., San Jose, CA) has commercialized the vapor phase resin adsorption technology. The Purus system is regenerated with nitrogen gas in a heated vessel. The recovered vapors are subsequently chilled prior to collection for disposal. A system schematic of the

Purus system is shown in Figure 9, and a photograph of the system is shown in Figure 10. These resins tend to adsorb VOCs in a fashion similar to GAC, but they also tend to demonstrate some absorption within the resin itself (Gusler et al., 1993). The regeneration capability of absorbed VOCs currently is not well quantified. Despite this uncertainty, the resins appear to have a few advantages over vapor phase GAC. The resin has a higher adsorption capacity for some VOCs than does GAC, is less influenced by relative humidity than GAC, and its on-site regeneration tends to produce less acid when adsorbing chlorinated VOCs than does GAC. The resin itself, however, appears to have some catalytic ability, and therefore breaks down some of the halogenated VOCs to form the hydrochloric acid (but still less than GAC). The resin also has less oxidative capacity and is less prone to bed fires than GAC for adsorption of readily oxidizable compounds.

In particular, the resins appear to better adsorb vinyl chloride than conventional GAC systems (alumina with permanganate impregnation is another potential adsorbent for vinyl chloride. It can be used as a final scrubber after a GAC system). An additional advantage of the resin system over steam regenerable GAC is the elimination of steam disposal/treatment. Although prices have been declining for these systems, they are still higher than conventional regenerable GAC systems. The average price of the resin is $70/lb in comparison to $1/lb for GAC, if purchased in bulk; $3/lb in canister form. The resin's superior performance, however, for certain compounds and its minimal disposal and O&M costs may make it more cost effective than regenerable GAC for certain installations. In general, the resin adsorption technology is not considered competitive for hydrocarbons, since these compounds can be more economically destroyed by catalytic or thermal oxidation. The resin technology is most applicable for moderately contaminated halogenated VOC vapor streams.

Oxidation-Based Technologies

Oxidation-based technologies react the VOCs at elevated temperatures with oxygen for sufficient time to initiate the oxidation reactions. The ultimate goal of any combustion/oxidation reaction is the conversion of the VOCs to carbon dioxide, water, sulfur dioxide, and nitrogen dioxide (end products of combustion). The mechanism involves rapid chain reactions which vary with the different VOCs being oxidized. The sequential reactions leading to combustion generally are not detrimental to the environment or to the oxidative equipment unless they are interrupted (possibly insufficient residence time or decayed catalysts). If the oxidation reactions are incomplete, partial oxidation byproducts will be released in the stack effluent. Sometimes these compounds may be more noxious than the parent compound. For example, oxidation of oxygenated organics may form carbon monoxide under unsatisfactory combustion conditions. To minimize such occurrences, excess air (above stoichiometric conditions) is used during oxidation reactions. The continuity of any oxidation reaction depends on maintaining the reaction mixture (air, VOCs, temperature, and catalyst, if used) in the optimal

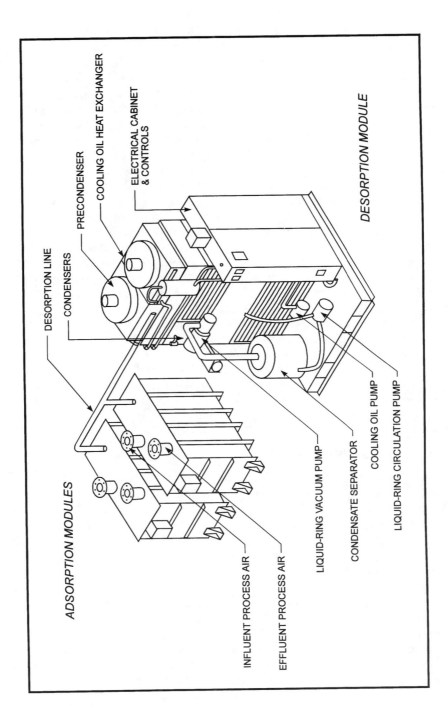

Figure 9 Schematic of the Purus resin adsorption system.

Figure 10 Dual-bed resin adsorption system. (Courtesy of Purus, Inc., San Jose, CA.)

mixture range (Bethea, 1978). The two most common oxidation methods in the remediation industry are thermal oxidation and catalytic oxidation. Figure 11 shows a schematic of the two methods.

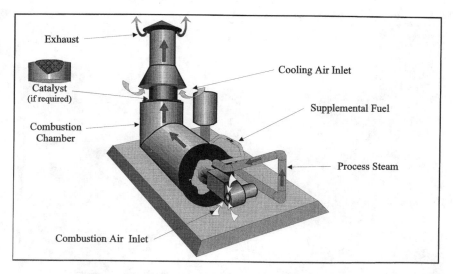

Figure 11 Trailer-mounted thermal and catalytic oxidizer.

Thermal Oxidation/Incineration

This technology burns the VOCs at an elevated temperature with resultant production of burn gases such as carbon dioxide and water vapor. Thermal oxidizers typically operate at 1400 to 1500°F with typical residence times of 0.75 to 2.0 seconds. The burner chamber must be designed so that the VOCs pass directly through its flame, thus minimizing the possibility of incomplete combustion. Although thermal oxidation is often the most expensive combustion-based control process, it is very well suited for high VOC air streams (2,000 to 10,000 ppm range). Operating costs can be offset to some extent if the gas stream is already considerably above ambient temperature by utilizing the combustion heat for preheating the air stream (heat recovery). The use of the technology for air streams that have significant BTU values also reduces fuel consumption. VOC destruction efficiencies of well-operated thermal oxidizers are generally in the range of 95 to 99% range. Thermal oxidizers generally have a maximum allowable concentration of influent VOCs (50% of lower explosion limit (LEL) or in the range of 7000 ppm (by weight) for most hydrocarbon applications. While it is technically possible to operate thermal oxidizers above the 50% LEL range, the increased temperature created by oxidation of process streams with concentrations higher than this often will exceed the temperature design criteria for most commercially available oxidizers (typically 1600 to 1700°F maximum.) During setup and operation, care must be taken to ensure that the free oxygen in the exhaust stream maintains at a minimum of 10%. Situations such as reducing conditions in the subsurface being vented or high influent concentrations leading to oxygen starvation by the oxidizer result in less than optimal destruction efficiency. Figure 12 shows a photograph of a commercially available thermal oxidizer.

Figure 12   Thermal oxidizer. (Courtesy of Thermtech, Kingwood, TX.)

Thermal oxidizers have been used for the treatment of high concentrations of hydrocarbon VOCs. If the influent concentrations exceed 13,000 ppm, dilution air is added. The oxidation of halogenated hydrocarbons may result in formation of byproducts and hydrochloric acid which may necessitate the treatment of the combustion gases. This limitation generally leads many practionors to use adsorption-based technologies rather than burn-based technologies for the treatment of halogenated VOCs. This technology is best suited for VES emission control at heavily contaminated sites. The most frequent application has been in the venting of free phase hydrocarbons atop the water table. Often times, thermal oxidizers are trailer mounted and are utilized during the initial phase of site remediation when VOC levels are high. As concentrations drop, power and fuel consumption generally rises dramatically, and the trailer-mounted unit is replaced by a more economical treatment method. As a general guideline, a 500-cfm thermal oxidation unit, with heat recovery capabilities, can be purchased for $60,000. Typical monthly fuel consumption for a 500 scfm, 5000-ppm VOC air stream is in the range of $1,500 to 1800, based upon a natural gas cost of $.40/therm (the standard unit of sale for natural gas, representing 100,000 BTU). As the concentration of this vapor stream decreases, the fuel consumption cost required to treat the process stream described above can increase to over $2500/month.

Catalytic Oxidation

Catalytic oxidation occurs when the contaminant-laden air stream is passed through a catalyst bed which promotes the oxidative destruction of the VOC to combustion gases. The presence of the catalyst bed allows for the oxidation to occur at a lower temperature than would be required for direct thermal oxidation. The primary advantage of catalytic oxidation is the decrease in supplemental fuel requirement (BTU not provided by the VOCs).

The catalyst metal surface must be large enough to provide sufficient active sites on which the reactions occur. Its surface must be kept free from dust or other noncombustible materials. Catalysts are subject to both physical and chemical deterioration. Physical deterioration results from mechanical attrition or overheating of the catalyst. Chemical deterioration most frequently is due to the presence of impurities in the VOC stream or from byproduct formation. For example, in VES applications of leaded gasoline, catalyst poisoning from the tetra-ethyl and tetra-methyl lead in the gasoline vapors will likely occur. Another form of catalyst deterioration is caused by exposure to halogens or sulfur-containing compounds. Halogen poisoning may occur from entrained water particles which contain chloride (especially in remedial applications involving salt water) or from chlorinated VOCs. Metals in entrained water particles may also act as poisons. Mercury, arsenic, bismuth, antimony, phosphorous, lead, zinc, and other heavy metals are common poisons. Lastly, the presence of high methane levels (either naturally occurring or escaping from pipelines) may cause catalyst damage.

Catalytic oxidation burns the VOCs at approximately 600°F for most hydrocarbon remediation applications utilizing a platinum or palladium catalyst. The technology is best suited for treatment of nonhalogenated hydrocarbons to avoid catalyst poisoning. Catalytic oxidizers generally are limited to maximum influent VOC levels in the range of 3500 ppm (since VOCs are a BTU source and the catalyst has an upper temperature limit). Higher influent concentrations will require dilution in order to reduce the influent concentration. Dilution, however, can reduce the extracted volume from the subsurface. A cost analysis is frequently used to justify operation in the catalytic mode using dilution vs. purchase of a combined catalytic/thermal oxidizer. As a general guideline, a 500-cfm catalytic oxidation unit, with heat recovery capabilities, can be purchased for $65,000. Typical fuel consumption per month for a 2000-ppm VOC, 500-cfm air stream is in the range of $900 to $1200 based upon a natural gas cost of $.40/therm.

Catalytic oxidizers are available as a stand-alone units with auxiliary heat sources as shown in Figure 13 or can be purchased as an internal combustion engine as shown in Figure 14. The combustion engine unit is attractive for pilot testing applications since it is a self-contained unit with its own fuel source. Internal combustion engine units however, tend to be less reliable, require more maintenance than conventional units, and are limited in capacity.

Figure 13   Trailer-mounted catalytic oxidizer. (Courtesy of Thermtech, Kingwood, TX.)

Catalytic oxidizers that can treat chlorinated VOCs are also commercially available. These units have not been cost effective and have thus not gained industry acceptance.

Biological Technologies

Biological treatment of air emissions has gained significant interest in Europe in the past several years. In the same fashion that vadose zone bioventing can be accomplished, it is envisioned that an above-ground reactor can be configured to

Figure 14   Internal combustion engine catalytic oxidizer. (Courtesy of VR Systems, Inc., Richmond, CA.)

degrade the VOCs. The above ground reactor is commonly referred to as a "biofilter". These units are now commercially available from several vendors. Figure 15 shows a photograph of a biofilter.

Figure 15   Biofilter. (Courtesy of EG & G Biofiltration, Pittsburgh, PA.)

This technology involves the vapor phase biotreatment of the air emissions by bacterial populations growing to fixed media. The fixed media (various combinations of plastic support media, compost, wood chips, and other) and water in the biofilter assist in the adsorbing/solubilizing of the VOCs for subsequent breakdown by the bacteria. Use of biofilters is relatively new in the U.S. but has gained widespread attention in Europe. Treatment efficiency is generally around 40 to 95% (compound and biofilter system specific). The more water soluble compounds generally are better treated than the nonsoluble compounds, since most of the degradation occurs in the water phase which is adsorbed to the support media.

Scrubbers

While most of this chapter has been devoted to organic compounds, some VES remediations may have inorganic compounds in the contaminated air stream. Hydrogen sulfide is probably the most widely found inorganic contaminant. Some sites have been known to have up to 30,000 ppm of hydrogen sulfide in the extracted air stream.

Chemical scrubbers can be used to remove the H_2S from the air stream. The scrubber looks very similar to an air stripper but operates in reverse. The air/water contact serves to remove the contaminants from the air and transfer them to the water stream. The water can then be treated with caustic and an oxidizing agent (hydrogen peroxide, ozone, supersaturated oxygen) to destroy the hydrogen sulfide. The high pH increases the rate of H_2S transfer, and the oxidizing agent hydrolyzes the H_2S. This rapid reaction tends to maximize the driving gradient from the air to the water. In some instances (low concentrations, 50 ppm range) the H_2S in air may be treated biologically rather than oxidatively.

Technology Selection Summary

Vapor emissions from site remediation should be evaluated on a case-by-case basis, depending upon the available site characterization data, regulatory requirements, life cycle considerations, and other project-specific data as previously outlined. Some generalizations, however, can be formulated regarding technology selection. Thermal oxidation is generally best suited for low air flow (air is expensive to heat) and high hydrocarbon VOC situations, such as VES emissions at sites impacted by NAPL. Catalytic oxidation is best utilized at hydrocarbon sites with moderate VOC impacts and low to moderate air flows. This can be found at a typical service station site without NAPL but high to moderate soil contamination. Off-site regenerable GAC is best utilized in instances where the mass loading is low. Steam regenerable GAC is best applied at high air-flow situations (no need to heat the air) and moderate VOC concentrations (low regeneration frequency). Steam regenerable GAC is applicable for most VOCs but its isotherms should be checked for the individual compound in order to ensure that regeneration frequency is acceptable. Resin-adsorption systems are best utilized for halogenated VOC air streams with moderate contamination. Biofilters are best suited for the treatment of low to moderate concentrations of soluble and biodegradable VOCs such as BTEX and ketones at low to moderate air flows. Biofilters tend to be bulky and therefore large mass loadings require very large units. Table 3 provides these general guidelines in a tabular format.

Table 3   Air Treatment Technology Selection Guidelines

Technology	Most applicable VOCs	Most applicable air flow	Most applicable mass loadings
Off-site regenerable GAC	HC, halogenated	All	Low
On-site regenerable GAC	HC, halogenated	All	Moderate
Regenerable resin	Halogenated	All	Moderate
Catalytic oxidation	HC	Low to moderate	Moderate
Thermal oxidation	HC	Low to moderate	High
Biofilters	Soluble HC	Low to moderate	Low to moderate

Note: HC = Hydrocarbons. Low air flow considered in the 100 cfm range. Moderate air flow considered in the 500 cfm range. High air flow considered greater than 2000 cfm. Low mass loading considered in the 5 lb/day range. Moderate mass loading is considered in the 50 lb/day range. Moderate mass loading is considered greater than the 100 lb/day range. Reader is advised to use this table as a general guideline ONLY.

REFERENCES

Bethea, R.M. *Air Pollution Control Technology,* New York: Van Nostrand Reinhold, 1978.

Brusseau, M.L., Jessup, R.I., and Rao, P.S.C. "Transport of Organic Chemicals by Gas Advection in Structured or Hetrogeneous Porous Media; Development of a Model and Application to Column Experiments," *Water Resourc. Res.,* vol. 27, p. 3189, 1991.

EPA, Soil Sampling and Analysis for Volatile Organic Compounds. Groundwater Issue. EPA/540/4-91/001 or NTIS DE91 016758, Office of Solid Waste and Emergency Response, U.S. Environmental Protection Agency, 1991.

Faust, S.D. and Aly, O.M. *Adsorption Processes for Water Treatment,* Stoneham, MA: Butterworth, 1987.

Gusler, G.M., Browne, T.E., and Cohen, Y. "Sorption of Organics from Aqueous Solution Onto Polymeric Resins." *Ind. Eng. Chem. Res.,* vol. 32, no. 11, p. 2728, 1993.

Noll, K.E., Vassilios, G., and Hou, W.S. *Adsorption Technology for Air and Water Pollution Control,* Ann Arbor, MI: Lewis Publishers, 1992.

Rixey, W.G. and King, J.C. "Wetting and Adsorption Properties of Hydrophobic Macroreticular Polymeric Adsorbents." *J. Colloid Interface Sci.,* vol. 13, no. 2, p. 320, 1989.

Weber, W.J. *Physicochemical Processes for Water Quality Control,* New York: John Wiley & Sons, 1972.

Yao, C. and Tien, C. "Wetting of Polymeric Adsorbents and its Effect on Adsorption Behavior of Hydrophobic Resin Particles." *Res. Polymers,* vol. 13, p. 121, 1990.

8

FRACTURING

Donald F. Kidd

INTRODUCTION

Regardless of the carrier fluid, low permeability, fine-grained soils and rock represent a significant challenge to *in situ* contaminant remediation alternatives. We have already discussed the problems of moving air and water carriers through these types of geologic conditions. Without the movement of these carrier fluids, *in situ* remediation methods are severely limited in their effectiveness. Despite the low permeability of clays, silts, and competent rock, these geologic formations can still become impacted. Over time, organic contaminants can permeate throughout a wide area of the subsurface in vapor, non-aqueous (NAPL), and aqueous phases migrating through natural fractures and by diffusion into the fine-grained soils. Once in these zones, rapid or sufficient removal of the contaminants is difficult to achieve, if possible at all.

Within low permeability settings, excavation and above-ground treatment/disposal or encapsulation are commonly selected remedies. As with all above-ground remediation, the excavation process may actually enhance the potential exposure of the population to the subsurface contaminants during the process. This is always an objectionable consequence of the cleanup process. The excavation process is also disruptive to ongoing facility operations, and impacted soil transported to a landfill poses some long-term liability.

A new technology, fracturing for permeability enhancement, is rapidly being developed to address these low permeability zones. The limitations on achievable contaminant reduction *in situ* are mainly due to inadequate carrier fluid exchange frequency or nonuniform distribution of the carrier fluid. The fracturing process seeks to increase soil permeability within discrete zones through the production of high permeability fractures. Both hydraulic and pneumatic fracturing are designed with this purpose in mind.

0-87371-995-6/96/$0.00+$.50
© 1996 by CRC Press, Inc.

APPLICABILITY

Almost any rock or soil formation can be fractured, given enough time, energy, and determination. The key aspect that has to be considered for remediation purposes are "Will the benefit derived from fracturing off-set the cost of the process?" and "What are the risks and benefits of the process?" Armed with the answers to these two questions, the decision to proceed with testing and, ultimately, full-scale application of the technique can be made on an informed basis.

Fracturing is most appropriately applied to soils where the natural permeability is insufficient to allow adequate carrier movement to achieve project objectives in the desired time frame. The following soil types and rock are generally treatable with the fracturing technologies (Schuring and Chan, 1993):

- Silty clay/clayey silt
- Sandy silt/silty sand
- Clayey sand
- Sandstone
- Siltstone
- Limestone
- Shale

Fracturing a sand or gravel formation, while possible, is probably not justified because the increase in soil permeability likely would be incremental.

Fracturing technologies are equally applicable to both vadose zone (unsaturated) soils and saturated soils within an aquifer. The idea is to improve the flow of carrier fluids for contaminant removal or delivery of nutrients or reactive agents.

By itself, fracturing is not a remediation process. There is no inherent advantage to having contaminants in contact with a high permeability formation, and, in fact, there can be disadvantages to this situation. Fracturing has to be combined with some other technology to be of benefit for reducing contaminant concentration, mobility or both. Essentially, fracturing serves solely to engineer changes in the subsurface so that carriers can more effectively reach the contaminants of concern. Contaminant removal and encapsulation processes by degradation, volatilization, dissolution (leaching), and stabilization are still controlled by the characteristics of the contaminants and the impacted media.

The important process of diffusion has been previously discussed in terms of how it impacts the spreading of contamination in the subsurface and the implications on the cleanup process. Controlling or limiting the role of diffusion on remediation is perhaps the most promising aspect of induced-fracture formation. The importance of this understanding cannot be over-emphasized to those responsible for developing remediation programs. Chapter 1 describes diffusion as a process by which dissolved chemicals move independently of the primary, advective flow path of impacted water or soil vapors. The chemicals can move into materials of low flow or even no-flow ("stagnant") conditions. Diffusion-based flow occurs due to molecular movement and is enhanced by differences in dissolved concentrations between areas of high flow into the relatively clean, low flow materials. When trying to reverse the contamination process (cleaning the impacted aquifer or vadose zone), these areas of diffusion-created pockets of contaminants can significantly extend the life of a project.

By fracturing, not only do we create higher permeability zones for enhancement of advective flow through the impacted material, we also shorten the pathway for diffusion-controlled flow of the carrier fluid. The creation of advective flow channels and shortened pathways for the lower velocity diffusive flow result in enhancement of the carrier delivery (or recovery) process. The final cleanup level attainable by fracturing and the associated remediation process still will be governed by characteristics of the soil/rock and contaminant. Diffusion-limited extraction still will influence the rate of contaminant recovery even after fracturing, and contaminant/media attractive forces still will influence the final concentration. This is important to understand.

As discussed in Chapter 4, the flow volume of the carrier fluid has a significant impact upon the rate of contaminant removal. The flow volume of the carrier is in turn a function of soil permeability and the *in situ* pressure gradient (pressure values between points of reference). The vapor velocity in the subsurface is then limited by the achievable volumetric flow rate at the extraction well. The following equation can be used to approximate this volumetric flow rate, given knowledge or estimation of soil permeability and radial influence (Johnson et al., 1990):

$$Q^*/H = \pi * (k/\mu) * P_w * [1-(P_{atm}/P_w)^2]/\ln(R_w/R_I)$$

where:

Q^*	=	flow rate (ft^3/min)
H	=	screen length (ft)
k	=	soil permeability (ft^2 or darcy)
P_w	=	well pressure (atm)
P_{atm}	=	atmospheric pressure (atm)
R_w	=	well radius (ft)
R_I	=	radial influence (ft)

Figure 1 illustrates the relationship between pressure gradient (applied vacuum levels), sediment permeability, and vapor withdrawal rates (Johnson et al., 1990).

The vapor flow velocity diminishes rapidly with distance from the point of extraction, again as discussed in Chapter 4. This occurrence results from expansion of the area through which the vapors pass. Mathematically, the vapor velocity through the soil or rock is calculated as flow (Q) divided by area (A). As an example of this relationship, Figure 2 illustrates a typical extraction well constructed with a 10-ft screened section within a low permeability clayey sand (k = 0.1 darcy). As illustrated on this figure, the area of flow is defined as:

$$A = 2 * \pi * r * H$$

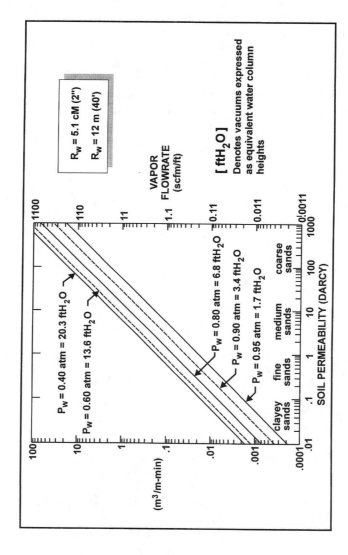

Figure 1 Pressure gradient *vs.* sediment permeability. (From Johnson, P.C. et al., *Groundwater*, vol. 28, no. 3, 1990. With permission.)

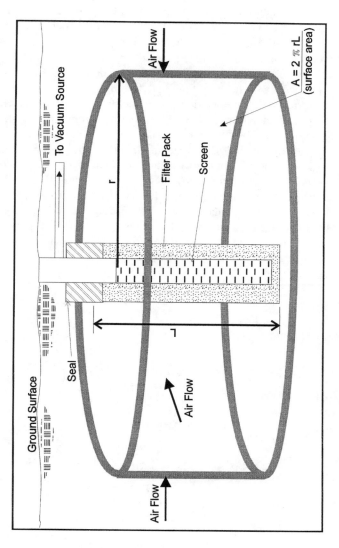

Figure 2 Illustration of radial flow and area of flow.

where

A = area (ft²)
r = radial distance (ft)
H = screen length (ft)

Note this estimation of flow area assumes that flow into the extraction well is primarily horizontal such that $k_v \ll k_h$. (K_v is vertical permeability and K_h is horizontal permeability). This approximation is especially valid for materials with a significant fraction of fine-grained particles.

For our hypothetical situation, an applied vacuum of 12 inches of mercury (inHg.) is predicted to result in a flow of 1.1 scfm based on the above equation. Figure 3 illustrates the predicted vapor velocity vs. radial distance from the extraction well, again assuming that the vertical flow component is negligible. A horizontal line is also placed at approximately 0.01 ft/min which represents the minimum "critical velocity" described in Chapter 4. In summary, this critical velocity represents the minimum recommended velocity to optimize contaminant recovery rates. As shown on Figure 3, for a clayey sand, the critical velocity occurs at distances less than 2 ft from the extraction well. To maintain vapor velocity in excess of the critical value, well spacings would be very tight. The benefit of fracturing in the situation described above is that more vapor can be withdrawn from the subsurface and the desired velocity profile can be extended farther out into the contaminated formation.

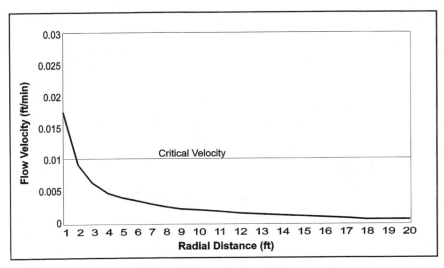

Figure 3   Flow velocity vs. distance.

Fracturing can also expand the applicability of other *in situ* remedial technologies beyond vapor and liquid extraction. As an example, hydraulic or pneumatic fracturing of a low permeability vadose zone overlying a more transmissive geologic unit can allow the use of air sparging (detailed in Chapter 6) in a geologic setting which is normally unsuitable. Generally, the injection of air beneath a low permeability formation can result initially in organic-laden vapor accumulation beneath this zone and eventually the lateral migration of these vapors. With the uncontrolled migration of contaminant vapors beyond the influence of a collection system, errant emissions can result, leading to unforeseen exposure routes or the contaminants re-entering the ground water through dissolution. The latter case would result in the expansion of the dissolved plume initially targeted by the remedial action. By installing a fracture network above the zone of aeration (sparging), the vapor collection system can recover the stripped contaminants, thereby avoiding these two undesirable occurrences.

Figure 4 provides a conceptual illustration of this application of fracturing. For the case illustrated, the formation overlying the impacted water producing sand zone is fractured and connected to a vapor collection system. Through the fractures, the vapors containing elevated contaminant levels are provided a capture zone.

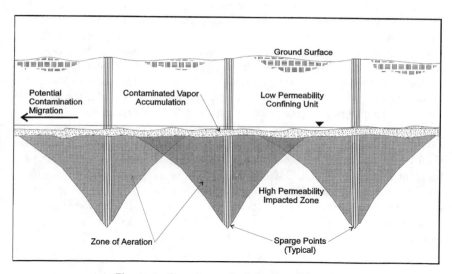

Figure 4   Sparging under low permeable soil.

Geologic Conditions

As with all remedial techniques, fracturing is only beneficial for environmental remediation for certain site conditions. In addition to the consideration of soil and rock types described in the previous section, the mode of deposition and changes occurring after deposition affect the effectiveness of fracturing. Most notably, the state of *in situ* stresses has long been characterized as the primary variable in the orientation of fracture formation (Hubert, 1957).

When fractures are formed by the injection of fluids, they are oriented perpendicular to the axis of least principal stress, with propagation following the path of least resistance. For environmental remediation, horizontal fractures are of the greatest benefit. Vertically-oriented fractures offer limited additional benefit to remediation as the fractures will tend to reach the ground surface at a relatively short distance from the injection point. Normally consolidated formations or fill materials have been found to produce vertically-oriented fractures. For vapor extraction technologies, described in detail in Chapters 4, the short-circuiting of vapor flow resulting from vertical fractures would actually be detrimental to the cleanup effort. Essentially, the soil vapor will follow the path of least resistance through the fractured media with little influence occurring beyond these engineered, preferential pathways.

In situ stress fields are subdivided into horizontal (x and y direction) and vertical (z direction) components. When initially deposited, sedimentary formations represent essentially hydrostatic conditions whereby the three principal stresses are in equilibrium and are equal to the weight of overburden. External forces (tectonics, burial/excavation, glaciation, and cycles of desiccation/wetting) after deposition then modify these stress fields.

Over-consolidation is defined as compaction of sedimentary materials exceeding that which was achieved by the existing overburden. Again, changes to the *in situ* stress fields after deposition have imparted a residual stress component to the formations. Over-consolidation of soils specifically results in stress fields favorable for fracturing. In this instance, the least principal stress is in the vertical direction. The induced fractures again would be created perpendicular to this stress and would be horizontally oriented. Figure 5 illustrates the concept of stress fields showing both equal and unequal stresses and the resulting orientation of fractures.

The formation and later retreat of glaciers is one condition which results in over-consolidation. The weight of the ice on the soil initially compacts the sedimentary grains. When the ice melts, the vertical stress is relaxed but the horizontal stress still maintains a residual component of the loaded conditions. This is not the only condition that results in over-consolidation, however. Erosion or overburden removal by excavation also present conditions which relax the vertical stress field. Additionally, the cyclic swelling and desiccation of clay-rich formations can also create conditions of over-consolidation.

TECHNOLOGY DESCRIPTION

With the current state of the technology described in this chapter, there are two types of fracturing methodologies employed for environmental applications. Hydraulic (water-based) and pneumatic (air-based) fracturing variants of permeability enhancement are described in the following sections. The selection between these two types of fracturing are based on these considerations:

- Soil structure and stress fields
- Contractor availability
- Target depth
- Desired areal influence
- Acceptability of fluid injection by regulatory agencies

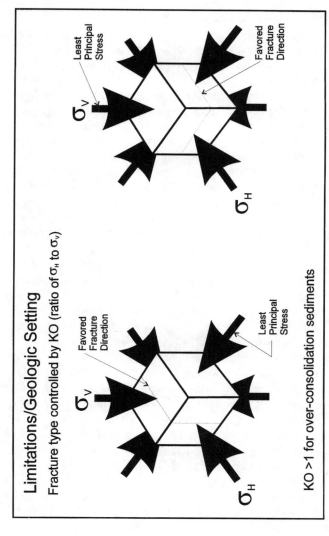

Figure 5 Pneumatic/hydraulic fracturing. (From Hubbert, M.K., *Petroleum Trans. AIME*, vol. 210, pp. 153–166, 1967. With permission.)

HYDRAULIC FRACTURING

Hydraulic fracturing was first developed as a means of enhancing oil and gas production. The first successful fractures completed for this purpose are credited to the Hugoton gas field in Grant County, Kansas, in 1947 (Gidley et. al., 1989). Early fracturing fluids were a gasoline-based, napalm gel and contributed significantly to the hazards of fracture installation. Since its beginning, over 1 million fracture treatments have been completed. Currently, 35 to 40% of all production wells are fractured to enhance production rates. The process is reportedly responsible for making 25 to 30% of U.S. oil reserves economically viable. In other words, many oil and gas reserves would not be produced with only naturally-occurring pressure distributions and gravity drainage-controlling recovery rates. The parallels between economic recovery of petroleum hydrocarbons and viability of *in situ* treatment alternatives are very evident.

As the names imply, the primary difference between hydraulic and pneumatic fracturing for permeability enhancement is the choice of penetrating fluids used by each technology to create the subsurface fractures. Hydraulic fracturing fluids are characteristically viscous, produce minimal fluid losses to the formation, and have good posttreatment breakdown characteristics. High viscosity is desirable for the fluids to create a wide fracture and transport the proppants into the formation. Low fluid loss is important to minimize the volume of injected fluid while achieving the desired penetration. Finally, posttreatment breakdown is necessary such that the injected fluids do not "clog" the formation. Cross-linked guar is an example of a common fracture fluid used for both petroleum reservoir stimulation and for environmental applications. This fluid is a common thickener used in the food production industry which essentially breaks down to water with very little residual materials deposited into the formation. A food-grade carrier fluid minimizes the potential for regulatory objections to the process.

Because of the characteristic high viscosity, fracture fluids are capable of transporting particles (termed propping agents or "proppants") through the fractures and out into the formation. These proppants then support the fractures upon relaxation of the injection pressure and, to some degree, prevent closure of the fractures. Silica sand is most commonly used in both environmental and petroleum applications due to its relatively low expense, range of particle size, and general availability. The use and applicability of proppants other than sand are described in the section on proppants later in this chapter.

Hydraulic fracturing is a sequenced process in which multiple fractures can be generated within the impacted soil or rock formation. The separation between fractures is dependent upon an economical evaluation and the physical characteristics of the soil. The desired result of the fracturing process is a formation that allows for the effective delivery of carrier fluids and results in a more rapid reduction of contaminant concentration, minimization of project costs, or, ideally, both of these occurrences. The mechanics of the fracturing process are described in a later section.

Pneumatic Fracturing

Fracturing of soil or rock formations can also be accomplished using a compressed air or other gas source. As with the hydraulic variant, pneumatic fracturing proceeds by isolating discrete zones of the formation and applying energy (in this case compressed gas). Inflatable packers with delivery nozzles within the isolated intervals of the formation are typically used (Figure 6).

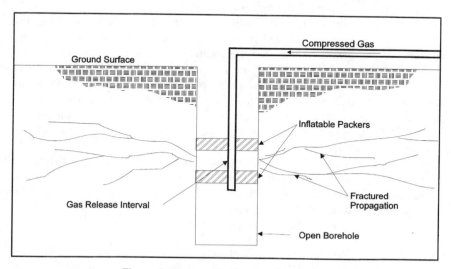

Figure 6 Pneumatic fracturing schematic.

To create the fractures pneumatically, compressed air is supplied at a pressure and flow that exceed both the *in situ* stresses and the permeability of the material. This energy then fractures the material and creates conductive channels radiating from the point of injection (Schuring et al., 1991). Injection pressures on the order of 150 pounds per square inch gauge (psig) and flow rates as high as 800 scfm or higher are used to create the fractures.

The pneumatic fracturing procedure typically does not include the intentional deposition of foreign proppants to maintain fracture stability. The created fractures are thought to be "self-propping". Essentially, disruption of the soil or rock structure during the injection of pressure results in localized realignment of grains within the fracture, preventing closure after relaxation of the pressure. Testing to date has confirmed fracture viability in excess of 2 yrs, although the longevity is expected to be highly site specific (Schuring and Chan, 1993).

Without the carrier fluids used in hydraulic fracturing, there are no concerns with fluid breakdown characteristics for pneumatic fracturing. There is also the potential for higher permeabilities within the fractures created pneumatically (compared to hydraulic fractures) as these are essentially air space and devoid of proppants.

SCREENING TOOLS

Fracturing success is dependent on the application of both sound engineering and sound judgment. The database of cleanup sites for which fracturing has been applied for testing and more so for full-scale remediation is very limited. With continued testing and reporting of both successes and failures, our understanding of the technology will develop to the point where geologic conditions favoring the technology will become better understood.

Screening a site for possible application of fracturing first requires that the project team not only understand the mechanics and applicability of fracturing to enhance permeability, but also the implications for the ultimate cleanup technology. By the point in the project when remedial alternatives are being considered, the extent of the contaminant impact should be defined. From this information, a preliminary estimate of the number of fractures necessary to provide coverage can be assessed from previous case studies. As a general rule of thumb, fracture formation in the range of 20 to 35 ft or more is possible for near-surface soils and, with all other factors remaining the same, increases with depth of burial. The relationship between burial depth (loading) and fracture dimensions needs to be considered for a full-scale application. Specifically, more closely-spaced shallow fractures may need to be created to achieve the desired end result. Fracture propagation in rock formations has been found to be greater than in soil formations, primarily due to the competence and cohesion of these units.

Knowing the limitations of the technology may result in its early "disqualification" based on site conditions and proposed objectives. As with any technology, the timely elimination of an alternative on a sound basis may result in overall project cost reductions and heightened focus on the final remedy.

Geologic Characterization

A primary step in the evaluation of fracturing applicability is an examination of detailed and accurate geologic cross sections illustrating sediment layering and the relationship between contaminants (target zone) and the different grain sizes. Because contaminants often reside within low permeability, fine-grained soils, this relationship is important to understand.

At least one continuous core boring should be installed during the remedial investigation phase of the project to characterize minor changes in lithology. Cores collected during continuous and depth-specific sampling should also be examined for factors contributing to secondary permeability (i.e., coarse-grained sediment inclusions, bioturbations, and naturally-occurring fractures). These secondary permeability characteristics of the soil or rock formation may influence the creation of engineered fractures. Pneumatic fractures, in particular, may propagate along existing fracture patterns. Hydraulic fractures have been found to be less influenced by existing fractures (Murdoch, 1993).

Geotechnical Evaluations

In addition to qualitative site evaluations described above, target zone soil samples can be submitted for geotechnical evaluations of grain-size analysis, Atterberg limits (plasticity and liquid limit testing), moisture content, cohesion, and stress evaluations (unconfined compressive strength). Details and implications of these tests are as follows:

- *Grain size analysis:* Although fractures can be created in sediments and rock of nearly any grain size, the highest degree of permeability improvement can be expected from the finer grained soils.
- *Permeability:* As discussed previously, fracturing is generally applied at sites with characteristically low permeability. A baseline estimate of permeability (vapor and/or liquid) is often available from testing conducted at the site during site investigation. This baseline estimate of permeability provides a basis of for evaluating the necessity, benefit, and effectiveness of the fracturing process. In general, greater improvement of carrier fluid flow or radial influence (in terms of percentage) is observed in formations with lower initial permeability.
- *Atterberg limits:* This parameter characterizes the plasticity of a soil. In general, fractures created in highly-plastic clays will not propagate as well as in more brittle materials. Formations having $W_n < W_l$ are most suitable for artificial fracturing, where W_n is the natural moisture content and W_l is the liquid limit. Soils having $W_n > W_l$ (or liquidity index greater than 0) may liquefy under a sudden shock imparted during the fracturing process. The estimation of W_n and W_p (plastic limit) would also give an indication of the degree of consolidation of soil. If W_n is closer to W_p than to W_l, the soil may be over consolidated. If W_n is closer to W_l (or larger), the soil may be normally consolidated.
- *Moisture content:* Overall soil permeability improvements are achievable with fracturing; however, vapor flow in particular is also controlled by soil moisture. Improvements in vapor flow through highly saturated soils (at or near field capacity) will not be achieved by the production of fracturing alone. Additional means of moisture removal may be required to obtain the desired effect from fracturing in these instances.
- *Cohesion:* Generally, the more cohesive the soil, the more amenable to fracturing and, upon relaxation of fracture stresses, longevity of the fractures. Fracturing in cohesive soils, such as silty clays, has been particularly successful.
- *Unconfined compressive strength:* The unconfined compressive strength can be used for predicting the orientation and direction of propagation of fractures. As evident from previous discussions, the

state of *in situ* stresses plays a key role in the orientation and ultimate utility of engineered permeability enhancement. The artificially induced fractures are assumed to be vertical in normally consolidated soil and horizontal in over-consolidated (or preconsolidated) deposits.

Pilot Testing

Upon completion of the preliminary screening and geotechnical testing, pilot testing is typically conducted for further performance evaluation and to provide a design basis for a full-scale system. Pilot testing is, by far the most powerful and useful means of screening a site for a full-scale remedial program incorporating fracturing. Experience has shown that preliminary screening of a site cannot always accurately predict the applicability or performance of either hydraulic or pneumatic fracturing.

To conduct pilot testing, it is wise to select a contractor who not only understands the applicability and limitations of fracturing, but also has experience in its specific field application. As the technology advances from infancy to a more mature status, the field of contractors meeting this criteria will increase along with the state of the art.

The pilot test plan should incorporate the following:

- Area selection
- Baseline permeability/mass recovery estimation (if not already done)
- Fracture point installation
- Test method and monitoring

Area Selection

Area selection of the cleanup site is the first step in designing the pilot program. The decision must be made whether to test the technology within the impacted area(s) of the site or conduct testing outside the contaminant zone. It is generally preferred to test within the area of impact to reduce the impact of lateral inhomogeneities and to provide data on contaminant recovery rates before and after fracturing. After all, the second criteria is the primary benchmark of fracturing performance: the influence and removal of contaminants.

For pilot testing of a single fracture well installation, an area of approximately 4000 to 8000 ft^2 should be sufficient. This area would encompass the anticipated maximum limits of the fracture propagation. Figure 7 provides a typical pilot test configuration. Note that both existing monitoring wells, as available, and specially-installed monitoring points are included within the test area. Testing procedures are briefly summarized below.

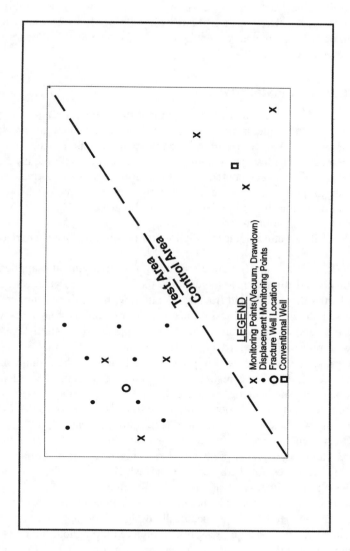

Figure 7 Pilot test configuration.

Baseline Permeability/Mass Recovery Estimation

To aid in the evaluation of fracturing benefits vs. the costs and risks of the technology, a baseline estimate of soil permeability and contaminant mass recovery rates are typically conducted prior to attempting to fracture the formation. Testing for permeability and mass recovery potential are described in detail in Chapters 4 and 5. Because fracturing is generally considered for low permeability formations (i.e., $k_{air} < 1$ darcy, $k_h < 10^{-4}$ cm/sec), the procedures necessary to evaluate these important parameters would most likely follow those outlined for vacuum-enhanced remediation.

Fracture Point Installations

A specifically-designed fracture point installation program is required for pilot testing of the fracturing technologies. The fracture intervals are selected to coincide with the known occurrence of the contaminants. Fracture locations are also targeted for the low permeability sediments or rock within a layered setting. The relationship between contaminant distribution and media permeability underscores the importance of a focused investigation program which provides this information during the assessment phase of the project.

With the variation between hydraulic and pneumatic fracturing, the field procedures for fracture point installation are briefly discussed separately.

Hydraulic Fracturing. Hydraulic fracturing for environmental applications initially involves drilling to near the depth of the fracture interval, advancing a lance to the depth of injection, and fracture initiation with a high pressure water nozzle (to create a notch with the desired horizontal orientation), followed by the controlled injection of the fracturing fluid and proppants. This multistage process is illustrated on Figure 8 (Murdoch, 1991).

Because the fractures will migrate naturally toward the ground surface, the volume of each injection is restricted to minimize this occurrence. At sites where hydraulic fracturing is being tested for the first time, this may present a trial-and-error sequence where the "optimum" injection volume is only discovered after daylighting of a near-surface fracture has occurred. "Daylighting" is defined as the observation of fracture media at ground surface. Continued injection of fracture fluids beyond this point will likely not result in additional subsurface fracturing. Once established, the fracture which has reached ground surface provides a relatively easy escape route for the fluids.

Injection pressures for hydraulic fracturing are typically less than 100 psig, with this pressure applied by pumps specifically designed for high viscosity, high solids fluid handling. The injection pressure is monitored throughout the process using a continuously-recording transducer or similar measuring device. An example of a pressure curve generated during fracturing is illustrated by Figure 9. As shown in this figure, the pressure initially builds up as injection is initiated. The pressure reaches a peak value and drops off dramatically. This drop in injection pressure signals the initial propagation of the fracture. Once this fracture is completed, the

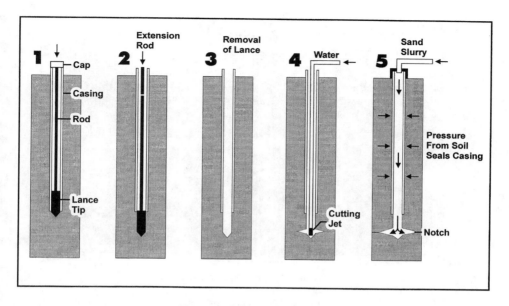

Figure 8 Hydraulic fracturing.

lance is withdrawn from the soil, the borehole is advanced, and the process is
repeated. Successive fractures are created in this manner and illustrated on Figure
10. The optimum separation between fracture intervals is site specific and best
determined during field testing, but is typically on the order of 2 to 5 ft.

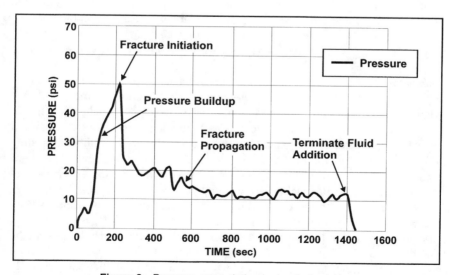

Figure 9 Pressure curve during hydraulic fracturing.

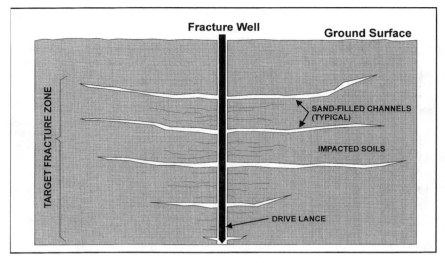

Figure 10   Successive fracturing.

The high viscosity fracture fluid must break down in place so that the benefit of a relatively high permeability zone can be realized. The commonly-used fracture fluid, guar gum, is often mixed with an enzyme to hasten the breakdown process although the guar will break down on its own, given sufficient time.

Upon reaching the desired maximum depth of fracture installation through the process described above, the borehole is often completed using convention well installation techniques. The placement of a well central to the point of radiating fractures allows for the withdrawal of vapors and liquids through the relatively permeable zones containing secondary permeability and proppants. Testing of the fracture performance is then conducted, similar to design-testing of vapor extraction, vacuum-enhanced recovery, or conventional aquifer analysis.

Pneumatic Fracturing. For the pneumatic variant of the fracturing process, a borehole is advanced to the desired depth of exploration and the augers are withdrawn. Variations on this technique may be required should noncohesive (e.g., flowing sands) be encountered. With the low permeability formation exposed within the borehole, a dual-interval packer is inserted and inflated, sealing off a discrete zone (Figure 6). With this zone isolated, a high-pressure, high-volume compressed air source is introduced between the packers. After completion of this fracture, the packers are deflated and moved to the next interval.

Compression of the borehole wall may occur during inflation of the packers. A tight seal must be obtained or the injected air will simply bypass the packers and vent through the borehole. With this compression, the borehole may not be suitable for use in the completion of a well as can be done with hydraulic fracture points. Carrier-fluid flows, passing through the damaged borehole, may actually be lower after fracturing than for a well completed in a more "standard" manner. The advantage gained during fracturing, therefore, would be eliminated.

Injection pressures as high as 150 psig and flows in excess of 500 scfm are used to initiate and propagate the fractures (Schuring and Chan, 1991) for pneumatic fractures. In contrast to the relative quiescent injection process for hydraulic fracturing occurring over a period from 10 to 60 min per fracture, pneumatic fractures are created over a period of approximately 20 seconds.

The moderately high flows and high pressures for pneumatic fracturing can be supplied by a bank of compressed air cylinders which are recharged between fractures. This configuration can meet the specific requirements of the process and still be relatively transportable: a criteria which has to be met for the economic viability of the technology. Compressors capable of this air delivery rate and pressure, while available in an industrial setting, are not applicable for pneumatic fracturing field work.

Test Method and Monitoring

Pilot testing of the fracturing technologies is generally a two-step process. The first step is conducted during the actual installation of the fractures during which time the approximate dimension and orientation of the fracture pattern is determined. The second step in the testing process is to determine the influence to carrier fluid movement within and beyond the area of fracture propagation and the corresponding contaminant removal rate improvement achieved through the process.

Fracture Orientation: Ground surface displacement (heave) is generally recorded during fracturing by an array of survey points which are monitored in real time. The change in ground surface elevation during fracturing has been found to provide a reasonable approximation of the fracture locations in the subsurface.

For pneumatic fracturing, the surface heave during pressure application is substantially higher than after pressure relaxation. The residual heave is generally 10 to 20% of the maximum displacement (typically less than a few inches) and provides evidence of fractures creation. For hydraulic fracturing, the ground displacement is directly related to the volume of proppant injected into the soil, with the displacement being a close approximation of the fracture dimensions (thickness of fracture zones). For this method of fracturing, proppant layers up to an inch thick have been created at each depth of injection. The thickness of the fractures diminishes with distance from the point of injection.

For both pneumatic and hydraulic fracturing, the fracture zones are generally asymmetric about the point of injection, following the path of least resistance. Heterogeneity within the soil matrix, naturally occurring fracture patterns, and, to a lesser degree, bedding planes appear to influence the orientation of the created fracture zones. Surface loading also influences the pathway of the fracture front. High surface loading created by manmade structures or changes in topography can also influence the fracture patterns. This tendency can be beneficial in some instances, as temporary surface loading can be used to "steer" the fractures toward the desired location. Vehicles have been used successfully for this application.

Because of the displacement caused by the fracture formation, care must be exercised when working adjacent to buildings or other structures. While some structures can withstand these moderate displacements, the integrity of others may be compromised. A careful evaluation of the strength and flexibility of the structure must be conducted prior to implementing a fracture program under circumstances such as these. This type of evaluation is beyond the scope of this text.

Carrier Fluid Influence: The dimension of fracture propagation is only one aspect of the pilot testing program and probably not the most important consideration. With the underlying and driving force behind the technology being the improvement of the efficiency of carrier fluid transport and fluid delivery, the second stage of a typical testing program is to measure and quantify how the transport has been modified. The enhanced flow characteristics are then compared with the baseline estimate of carrier fluid flow characteristics, as discussed earlier, and a determination is made as to the relative benefit of the process. With this key information, the applicability of the process and, ultimately, the number, locations, and depth intervals of a full-scale fracture program can be designed. As described in detail in previous chapters, the second phase of the pilot testing program may consist of a groundwater pumping test, vapor extraction test, or vacuum-enhanced pilot test.

PROPPANTS

In contrast to proppants utilized solely for maintenance of fracture aperture or viability, reactive or conductive agents also have been recently promoted due to their characteristic properties. For example, a "time-release" oxygen source was developed in cooperation with the Environmental Protection Agency (EPA) which consists primarily of sodium percarbonate (Vesper et al., 1993). This proppant, when injected into the impacted soils, reportedly will slowly release oxygen over a 4-month period. The advantage of this occurrence is that the aerobic conditions can be locally maintained in the soil without active vapor withdrawal or injection.

Similarly, testing is currently underway for the injection of iron-filings for the creation of subsurface conditions favoring dehalogenation of chlorinated solvents. A flat-lying "reactive wall" is thus created which testing indicates can promote the accelerated attenuation of chlorinated solvents such as Perchloroethylene (PCE) and Trichloroethylene (TCE) without the high cost of extraction, above-ground treatment, and disposal. The concepts behind reactive wall applications are presented in Chapter 9.

A graphite-based proppant is also being tested for the enhancement of electro-osmotic dewatering and *in situ* resistive heating due to its electroconductivity. This process, named the "Lasagna Process" by its developers creates a sequence of flat-lying conductive beds within the impacted soil or rock. With the cyclic application of low-voltage direct current to these beds, osmotic flow is induced within the impacted formation. An acid front is created within the treatment zone. The contaminants are desorbed and removed from the permeable fracture zones by vapor extraction or direct pumping. The Lasagna Process is also detailed in Chapter 10. While this technology may be very beneficial for relatively rapid contaminant removal from low permeability formations, it has not been tested under field conditions as of this publication.

FULL-SCALE DESIGN

Upon completion of site screening, and pilot testing, a full-scale fracturing program may be implemented. A full-scale fracturing program (beyond the installation of a few fracture points) should not be attempted without field verification of its performance. The implementation of the full-scale program would be based on economic and feasibility evaluations. In essence, fracturing would be selected as a component of a final remedy if the cost of fracture creation is less than what would be required for multiple well installations with lower well spacings or alternate strategies, such as excavation and above-ground treatment/disposal.

Based on the field testing, a fracture well pattern would be selected to encompass the area of known contaminant impact. This pattern must take the asymmetric orientation of the fracture propagation. For instance, the fractures may be more closely spaced north to south than east to west due to asymmetry. To control the role of diffusion and the possible creation of "low flow zones", an engineering safety factor should be applied such that the fractures overlap.

The depth intervals for the fractures should correspond to the known distribution of contaminants. This requirement again emphasizes the importance of site characterization.

Because of heterogeneity present at almost every site, the full-scale fracturing program should be designed with some flexibility in mind. In most instances, it would be wise to specify a range of possible fracture point placements with adjustments made during implementation of the program to optimize the performance of the fractures. For example, it may be necessary to tighten the fracture spacings and reduce fracture volume (for both pneumatic and hydraulic fracturing) if daylighting is found to occur in one or more areas of the site, even if pilot testing of the same parameters did not produce this result.

Depending on the size of the project and number of fracture points, it may also be advisable to implement the fracturing program in a phased approach. For example, fracture wells could be installed on a 1-week cycle. During the first week, fracture points could be installed, followed by testing of these points for performance (carrier fluid extraction and movement rates). Adjustments can then be made for the next cycle of fracture installations.

The limiting role of diffusion cannot be overemphasized. Even with fracturing, contaminant removal rates will be rate limited by diffusional flow between the areas of high, advection-controlled flow. When compared to contaminant removal rates before fracturing, post-fractured rates will be higher, if the process is successful and applied under the right conditions. Eventually, however, diffusion-controlled flow will again predominate. The impacted soils may not have reached the target concentration by this time.

CASE HISTORIES

Fracturing tests, and to a lesser degree, full-scale remediations have been conducted at a number of facilities for both saturated and unsaturated zone soils. The following section provides a brief synopsis of some case studies conducted under the direction of the Superfund Innovative Technology Evaluation (SITE) program sponsored by the EPA. The case studies illustrate the performance of both pneumatic and hydraulic fracturing.

Pneumatic Fracturing Air Phase

A site in New Jersey was selected for a demonstration of the effectiveness of pneumatic fracturing. The impacted horizon at this site is characterized as siltstone and shale with naturally occurring fractures. Fractures were installed between 9 and 16 ft below grade. The primary contaminant was TCE.

Before fracturing, the vapor extraction rates from each of the tested wells was below the sensitivity of the measuring instrument (<0.6 scfm) at an applied vacuum of 136 inH$_2$O. A single fracture well was installed central to the monitoring points, as illustrated on Figure 11. The distances between the fracture well and the monitoring points was 7.5 to 20 ft.

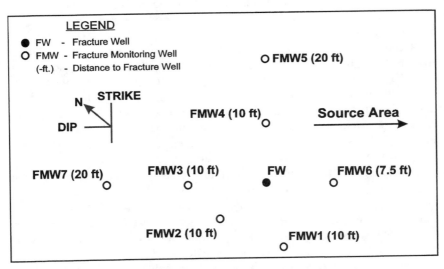

Figure 11 Case history — well location plan.

Based on elevation measurements recorded by an electronic tiltmeter during fracturing, surface heave was observed up to 35 ft from the fracturing well. The flow rates from each of the test wells surrounding the fracture well increased substantially after fracturing. Specifically, the flow rate increased from 3- to more than 15-fold after fracturing as illustrated in Figure 12. Vacuum measurements within the monitoring points also increased post-fracture from 4- to almost 100-fold. Contaminant recovery rates increased by a factor of approximately 8 to 25 as a result of the permeability enhancement.

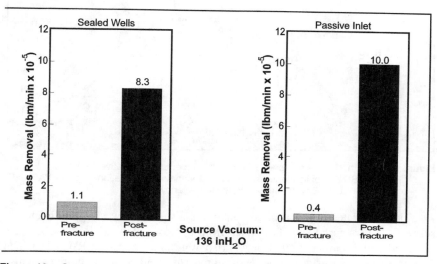

Figure 12 Case history — mass removal. (From Hillsborough, NJ EPA SITE Demonstration, 1993.)

A downhole video camera was used to observe the fracture patterns created in the formation. The inspection of the fracture well found that the primary mode of propagation was through existing fractures, although some new fractures were created.

EFFECTIVENESS — HYDRAULIC FRACTURING AIR PHASE

Hydraulic fracturing tests were conducted on vadose zone soils at a site in Illinois which were impacted with TCE, 1,1,1-TCA (Trichloroethane), 1,1-DCA (Dichloroethane), and PCE (Perchloroethane) (EPA, 1993). The soils are characterized as a silty/clayey to a depth of approximately 20 ft below grade. The conductivity of the soil was estimated at 10^{-7} to 10^{-8} cm/s. The pilot-scale demonstration created six fractures in two wells at depths of 6, 10, and 15 ft below grade over a 1-day period. A site plan showing well and monitoring point locations is provided in Figure 13.

At an applied vacuum in excess of 240 inH_2O, the vacuum influence in unfractured soil was negligible, decreasing to a few-tenths of an inch of water at a distance of 5 ft from RW-1 and RW-2, clearly demonstrating the limitations of a standard vapor extraction system under the site conditions.

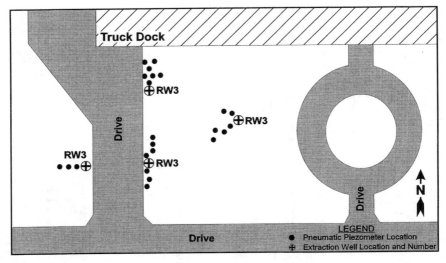

Figure 13 Case history — site plan. (From EPA, Superfund Innovative Technology Evaluation (SITE) Applications Analysis and Technology Evaluation Report EPA/540/R-
93/505, U.S. Environmental Protection Agency, 1993.)

The flow rates in unfractured soils were also very low at approximately 1
scfm. In the fractured soil, flow rates from approximately 14 to 23 scfm were
achieved under similar vacuum levels (isolating one interval of RW-4 in which
fractures vented to surface). Suction head measurements in the fractured soil also
increased dramatically up to 25 ft from the fracture wells. Contaminant recovery
rates similarly increased in fractured soil from 7 to 14 times after fracturing.

REFERENCES

EPA, "Hydraulic Fracturing Technology," EPA/540/R-93/505, Superfund Innovative Tech-
    nology Evaluation (SITE) Applications Analysis and Technology Evaluation Report,
    U.S. Environmental Protection Agency, 1993.
Gidley, J.L., Holditch, S.A., Nierode, D.E., and Veatch, Jr., R.W. *Recent Advances in
    Hydraulic Fracturing,* Richardson, TX: Henry L. Doherty Memorial Fund of AIME,
    Society of Petroleum Engineers, 1989.
Hubbert, M.K. "Mechanics of Hydraulic Fracturing," *Petroleum Trans. AIME,* vol 210, p.
    153–166, 1957.
Johnson, P.C., Kemblowski, M.W., and Colthart, J.D. "Quantitative Analysis for the
    Cleanup of Hydrocarbon-Contaminated Soils by In situ Soil Venting." *Groundwater,*
    vol. 28, no. 3, 1990.
Murdoch, L. "Feasibility of Hydraulic Fracturing of Soil to Improve Remedial Actions."
    EPA/600/S2-91/012, U.S. Environmental Protection Agency, 1991.
Murdoch, L. "Hydraulic and Impulse Fracturing Techniques to Enhance the Remediation
    of Low Permeability Soils," unpublished paper, 1993.

Schuring, J.R., Jurka, V., and Chan, P.C. "Pneumatic Fracturing to Remove VOCs." *Remediation,* Winter 1991/1992.

Schuring, J.R. and Chan, P.C. "Pneumatic Fracturing of Low Permeability Formations — Technology Status Paper," unpublished paper, 1993.

Vesper, S.J., Murdoch, L.C., Hayes, S., and Davis-Hooper, W.J. "Solid Oxygen Source for Bioremediation in Subsurface Soils." *J. Hazardous Mater.,* vol. 36, no. 3, p. 265–274, 1993.

9

REACTIVE WALLS

Peter L. Palmer

INTRODUCTION

Reactive walls are a developing technology that shows a lot of promise and is receiving a lot of attention and research dollars. In principal, a permeable wall containing the appropriate reactive material is placed across the path of a contaminant plume. As contaminated water moves through the wall, the contaminants are removed or degraded, allowing uncontaminated water to continue its natural course through the flow system. Much of the work on reactive walls has been performed by the Waterloo Center for Groundwater Research, University of Waterloo, which has successfully tested small-scale prototypes of various reactive walls at the Canadian Forces Base Borden in Ontario. A good summary of recent developments has been compiled by Gillham and Burris (1994), and some of the information contained herein has been excerpted from this paper.

Reactive walls are gaining a lot of attention, not necessarily because they speed up the remediation process, but because they recognize the limitations of ground water cleanup programs and factor these limitations into minimizing the life cycle costs of remedial programs. Reactive walls rely on the natural movement of water to carry the contaminants through the reactive wall, where they are removed or degraded. Reactive walls attempt to eliminate or at least minimize mechanical systems, thus minimizing long-term operation and maintenance costs that so often drive up the life cycle costs of remedial projects. Long-term operation and maintenance costs are reduced because the site does not need a continuous input in energy and manpower. Failures due to mechanical breakdowns are also reduced. In addition, technical and regulatory issues concerning discharge of treated ground water are avoided.

Since it is a developing technology, there are few commercial applications where the technology has been proven. Consequently, this chapter will focus on concepts, potential applications, and methodologies for installing and using reactive walls in remedial programs.

0-87371-995-6/96/$0.00+$.50
© 1996 by CRC Press, Inc.

GENERAL REACTIVE WALL DESIGNS

Reactor walls rely on ground water to carry the contaminant to the wall where it is removed or degraded. The walls can be constructed of a variety of materials or they may utilize natural geology. The wall area can be used to create chemical or biochemical reactions or simply to facilitate a process to remove the contaminant. In its simplest form, the reactive wall (curtain) is similar to that shown in plan view in Figure 1 and in cross section in Figure 2. As the reader can see in these figures, a plume is migrating downgradient from a source, and an *in situ* permeable wall (of natural geology or manmade material) is present to remediate the constituents in the plume *in situ*. For example, if the plume contained VOCs, the reaction curtain could be a series of air sparging points, which would transfer oxygen into the plume and rely on air to carry the contaminants vertically for release to the atmosphere or for capture by a vapor extraction system. This design could be used for degradable organics (oxygen transfer for the natural bacteria to degrade the compounds) or nondegradable volatile organics (air carrier to remove the compounds). This simple design would only be viable in geology that could facilitate air sparging.

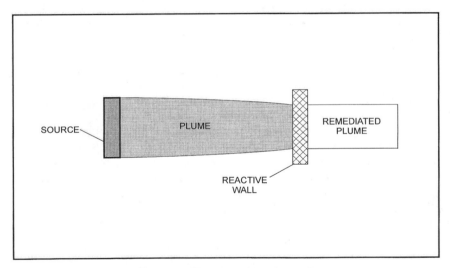

Figure 1   Plan view of reactive wall.

Many times the geology prevents the application of a technology or we need to add reactive material to the wall. In these cases, we must remove the natural geology and create a wall out of the required material. To continue with the above example, if the native materials are low-permeable sediments, then an additional purpose of the curtain could be to "change the geology" by excavating the native material and backfilling with more porous material which would be amenable to air sparging. Another method that is being studied for treating chlorinated hydrocarbons is the use of elemental iron (and other metals) to dehalogenate the

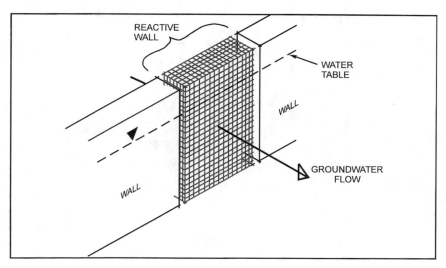

Figure 2 Cross section of reactive wall.

compounds. As will be discussed later in this chapter, the wall would then include elemental iron when it was constructed.

To successfully remediate a plume, a reactive wall must be large enough to remediate the entire plume. For large or deep plumes, this becomes impractical. To overcome this problem, a system can be installed consisting of low permeability cutoff walls, which funnel flow to a smaller reactive wall (referred to as a "gate") to treat the plume (Figure 3). This system is referred to as the "Funnel-and-Gate®" (Waterloo Groundwater Control Technologies, Inc., Waterloo, Ontario, Canada) system. There are any number of combinations or configurations which can be used to effectively control and remediate a plume, and these are discussed in detail by Starr and Cherry (1993). For instance, Figure 3 shows a single-gate system, and Figure 4 shows a multiple-gate system consisting of three gates. When dealing with funnel-and-gate systems, in all cases the sole purpose is to use the gate to pass contaminated ground water through the reactive wall which remediates the ground water. The funnel is integrated into the system to force water through the gates and is used for practical and economic reasons. Slurry walls, sheet piling, and other materials which could form the funnel are often times easier or more economical to install than the reaction wall. Consequently, the design is focused on balancing the ratio of funnel to gate areas to achieve remedial objectives at least cost. It must be remembered that ground water travels at a relatively low speed; therefore, the residence time in the reactive portion of the wall will be significant even when the reactive portion of the wall is not continuous.

Conceptually, plumes with a mixture of contaminants can be funneled through a gate with multiple reactive walls in series. For instance, one reactive wall could be used for degrading hydrocarbons and a second reactive wall in series could be used for precipitating metals as shown in the example in Figure 5. In

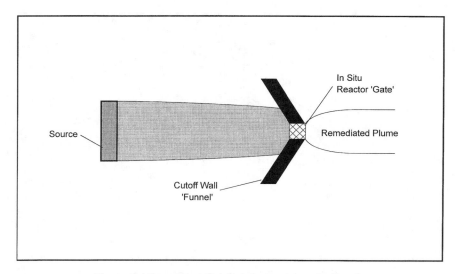

Figure 3   Funnel-and-Gate® system using a single gate.

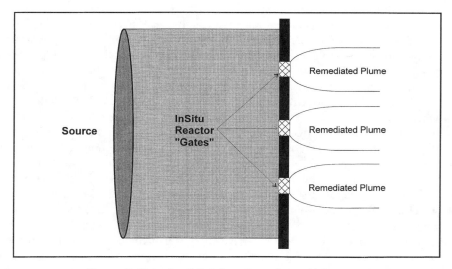

Figure 4   Funnel-and-Gate® system using multiple gates.

addition, if the gate needs to be removed at some point during or after remediation such as in the case with sorption processes (activated carbon, ion exchange, etc.) then considerations should be given to installing a retrievable wall, referred to as a "cassette". These could take the form of different shapes but they would each have sufficient permeability to allow migration of ground water through the container holding the reactive material, and the container would have sufficient strength to maintain its shape and structural integrity during placement and removal.

There are numerous different nuances to installing a reactive wall. For instance, Figure 6 shows a reactive wall designed to remediate a shallow plume that is located in the uppermost portion of the aquifer. Since the plume remains shallow, the reactive wall does not need to penetrate the entire thickness of the aquifer and is referred to as a "hanging wall (gate)". In situations, where a hanging wall is under consideration, it is important to understand contaminant transport to ensure that the contaminants remain in the upper portion of the aquifer, even after the subsurface is disturbed as a result of installation of the reactive wall.

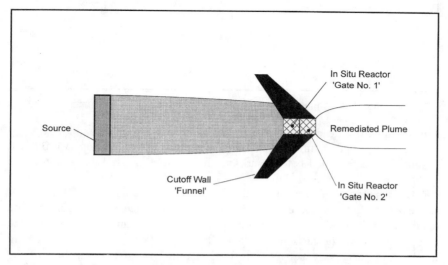

Figure 5 Funnel-and-Gate® system using a series of gates.

Installation Methodologies

There are a number of methodologies for installing reactive walls, including excavation and backfilling with permeable or reactive material, overlapping caissons, and soil mixing. Excavation and backfilling can be accomplished several different ways but they normally involve digging an open trench and stabilizing the side walls until after the trench has been backfilled with permeable or reactive material. Traditional methodologies such as steel sheet piling can be used whereby two rows of sheet piling are driven to the desired depth and the soil between them excavated; after placement of the wall material the sheet piling is removed. A biodegradable polymer slurry, which has been used in drilling water wells for decades, also can be used. Similar to a slurry wall, the biodegraded slurry would provide physical support to hold up the trench side walls during wall construction. Where it differs from a slurry wall is that after

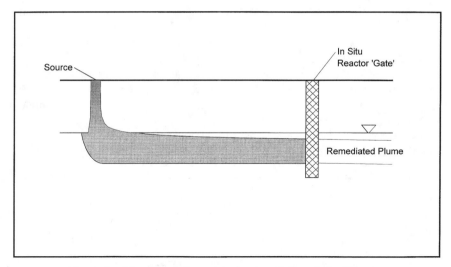

Figure 6 A hanging gate system for remediating shallow plumes.

wall completion the polymer is flushed out with the remainder biodegrading over a short period of time. This method is applicable to depths up to 100 ft and it should be determined beforehand that the reactive material and polymer are compatible.

Overlapping steel caissons are also applicable for depths up to 100 ft. Similar to steel sheet pilings, the caissons are driven or vibrated down to the desired depth; the soil within them is removed with an auger and replaced with the reactive material. The caissons are then withdrawn. Steel caissons do not require the bracings necessary between steel sheet pilings, and the selection of the preferred method probably would be based upon site constraints (available space, proximity to structures whose foundations could be sensitive to displacement, etc.) and costs.

Soil mixing has been used in the environmental field primarily for *in situ* mixing of solidification and stabilization agents to physically or chemically bind the contaminants of concern (primarily metals) to minimize their mobility. For reactive walls, a similar process would be used whereby large-diameter mixing augers are drilled into the subsurface, and during the process reactive wall additives are injected through the hollow stem. This method is also good to depths of 100 ft and is desirable because little or no soil is removed, minimizing soil management costs. The biggest concerns involve the ability to get complete mixing between the native soil and the reactive material. Of course, for this methodology to be effective, the native material must have sufficient permeability and must be compatible with the reactive material.

In each case, if the wall is installed within the plume, the excavated soil needs to be managed properly, and if it is designated as a hazardous waste, the cost of wall construction could be significantly increased.

TYPES OF *IN SITU* REACTORS

A variety of *in situ* reactors are being evaluated, and most are either in the treatability or pilot-study stage. The types of *in situ* reactors currently under evaluation include:

- Transformation processes
- Physical removal
- Modify pH or oxidation-reduction potential (Eh) conditions
- Precipitate metals
- Contaminant removal via sorption
- Nutrient addition

Presented below is a brief description of some of the processes under evaluation.

Transformation Processes

Recent studies at the Waterloo Center for Groundwater Research have focused on the use of zero-valent metals to promote transformation of various chlorinated organic compounds. Researchers have concentrated on using zero-valent iron to dehalogenate chlorinated aliphatic organic compounds both in the laboratory and in field tests (Gillham et al., 1993). In the laboratory they studied various metals including stainless steel, copper, brass, aluminum, iron, and zinc using 1,1,1-TCA, and found iron and zinc to be most effective. They focused their follow-up tests on a range of chlorinated organic compounds using only iron, because of lower costs and availability.

The results were very encouraging for all compounds tested (halogenated aliphatics including four methanes, four ethanes, and six ethanes), except for dichloromethane which showed no evidence of dehalogenation. Based on their studies, they offered the following:

- *Biological activity:* Although the tests confirmed the process was abiotic and biological activity does not affect the process directly, indirect effects such as biological coatings on active surfaces or biochemically induced changes in geochemistry could be factors in inhibiting the reaction rate during long-term operations.
- *Surface area to volume rates:* It was hypothesized that the reaction involves a surface phenomenon associated with the iron. As expected, the rate of degradation increased as the ratio of surface area of the iron to the volume of solution increased.
- *Effect of pH:* The rate of reaction appears to be a first order equation; however, the rate of reaction declines with increasing pH and is suspected to decrease with accumulation of reaction products.
- *Effect of dynamic flow conditions:* As a first-order reaction, the concentration profile shows an exponential decline from the initial concentration at the influent end. As the flow velocity increased, the profile became elongated.

Based on the encouraging laboratory tests, a field test was performed. The reactive wall was constructed into medium- to fine-grained sands by driving sheet piling to form a cell about 5 ft wide by 17 ft long. The wall was formed using 22% iron grindings and 78% concrete sand by weight with a permeability greater than the native sands. The sheet piling was then removed and a plume of Perchloroethylene (PCE) and Trichloroethylene (TCE) allowed to pass through the wall.

The test was conducted and monitored for 474 days, and there was no evidence in a decline in the effectiveness of the wall in degrading PCE and TCE. Figure 7 shows the maximum concentrations of TCE, PCE, and chlorides 299 days into the test. As shown, the wall resulted in significant mass transformation of TCE and PCE, particularly within about the first 1.5 ft (5 days residence time) of the wall. Although the wall dehalogenated approximately 90% of the mass, ±10% of the mass was not affected. It was concluded that the initial portion of the wall performed according to laboratory studies, but at greater distances into the wall performance was below expectations. The increase in pH is suspected to be the cause of the reduction in reaction rates. Increasing chloride concentrations shown in Figure 7 attest to the dechlorination of PCE and TCE, although the data was insufficient to conclude that dechlorination went to completion.

From the results of the laboratory and field test it was concluded that zero-valence metals, particularly zinc and iron, can significantly enhance the rate of degradation of many chlorinated aliphatic compounds due to reductive dehalogenation. The results also suggest that if increases in pH within the wall can be controlled that the wall may maintain its effectiveness throughout its entire width. Lastly, during the 14-month period, reactive wall performance was not adversely affected by precipitation or biological processes within the wall, providing encouraging results.

Many full-scale tests are now planned for this technology. The readers will have to rely upon published papers in order to keep abreast of the advances in this area.

Physical Removal

The best example of physical removal was discussed earlier whereby air is used to physically remove volatile organic compounds from ground water. This is an adaptation of air sparging; however, where it differs is that in a reactive wall, the geology is changed to increase soil permeability, in order to make sparging applicable. Using the funnel-and-gate concept, a funnel would be constructed to direct ground water flow to one or more gates and the gates would consist of materials more permeable than the native material, with air sparging points placed within to remove the volatile organic compounds from the ground water.

Modify pH or Eh Conditions

Modification of pH conditions occurs when the pH in the contaminant plume is raised or lowered. The main use of this application to date has been to precipitate

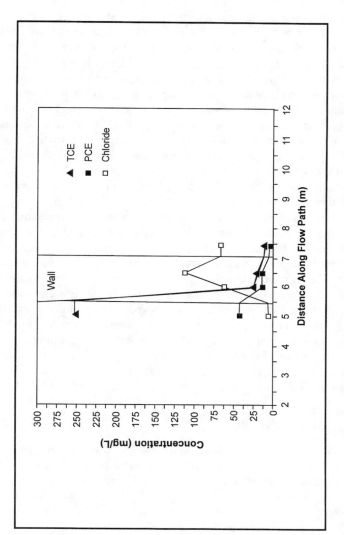

Figure 7 Concentrations of chlorinated compounds across the *in situ* treatment wall.

dissolved metals in a plume. However, this technology also could be used as part of a multiple wall application. The first wall could be used to create the necessary environment for the reaction in the second wall. A biological reaction zone may require a prewall to adjust the pH into a range in which the bacteria may grow. While organic destruction would be the main purpose of the design, the pH would have to be modified before the bacterial reaction could occur.

The main use of this application is for metal removal. For instance, at an abandoned mine site in the west, a plume of ground water with a pH less than 3.0 was discharging elevated concentrations of copper and zinc to surface waters. To solve the problem, a reactive wall was installed adjacent to the seepage face of the plume. The wall was designed to increase the pH of acidic ground water by conveying the ground water through a limestone bed to precipitate metal hydroxides. This application was designed to have a contact time of 1 h and has been operated successfully for over a year. Although this reactive wall deviates somewhat from a truly *in situ* reactive wall (it was constructed at the surface instead of below ground), it demonstrates the concept and suitability for using reactive walls for pH adjustments.

Modification of Eh involves the exchange of electrons between chemical species that effects a change in the valence state of the species involved. To be successful in reactive walls, the redox (oxidation-reduction) reaction must occur readily under the natural temperature and chemical setting of the ground water.

Blowes and Ptacek (1992) proposed to incorporate solid-phase additives as the reactive material in reactive walls to control redox conditions to promote removal of electroactive metals. Their work focused on using reduced iron (fine-grained, zero-valent) for removal of dissolved hexavalent chromium. Column studies in the lab demonstrated that highly reducing conditions generated by the reduced iron caused the chromium to change its state to the less soluble trivalent chromium which then precipitated (chromium hydroxide). A pilot test at the Coast Guard Air Station, Elizabeth City, North Carolina has been underway since September 1994 and the preliminary results show reductions of hexavalent chromium to below detection levels (Puls, R.W., Powell, R.M., and Paul, C.J., 1995).

Precipitation of Metals

Precipitation of metals could be accomplished by altering pH or Eh conditions as discussed above, or it could be accomplished by chemical additives to a reactive wall. For instance, in a plume containing dissolved lead, hydroapatite added to a reactive barrier could serve as a source of phosphate to precipitate lead phosphate. Studies performed by Schwartz and Xu (1992) demonstrate its potential applicability to reactive walls.

Other chemical reactions are possible. The key to the design for precipitation of metals is to have the necessary chemical in a solid phase form so that it can be part of the wall construction. The chemical still must be able to interact with the metal from the plume. Another example would be to use sulfide to precipitate metals. Metal sulfides generally are less soluble then metal hydroxides. Sulfides also will transfer from one precipitate to another. The wall could be constructed of calcium or iron sulfide. Metals that have lower solubility as sulfides (silver, mercury, etc.) would take the sulfide from the calcium or iron and precipitate. The calcium or iron then would be released to the water. Laboratory and pilot plant studies will be required before all of these new applications are ready for the field.

Contaminant Removal via Sorption or Ion Exchange

Contaminant removal via sorption also shows early promise but as with any developing technologies a number of issues need to be resolved. Sorption could be accomplished using a number of commonly used above-ground sorptive material such as activated carbon, zeolites, or ion exchange resins.

Activated carbon, for instance, is used to physically adsorb a variety of organic compounds, particularly synthetic organic contaminants in ground water. The activated carbon physically absorbs the organics due to the attraction caused by surface tension of the activated carbon, and it has gained widespread use in above-ground applications for ground water treatment because of the large internal surface area that activated carbon possesses. Since physical adsorption is a reversible process, its application in reactive walls requires that at some point, either after the carbon is spent or the plume is remediated, the carbon must be removed from the subsurface and managed accordingly. If the carbon is left below ground, it will release the adsorbed organics to the clean water carrier. This requires that the activated carbon be installed in retrievable containers (cassettes), discussed earlier in this chapter. Activated carbon reactive walls require consideration of other factors which could interfere with their adsorptive abilities. For instance, dissolved iron could be present which could compete with organics for adsorptive surfaces on the activated carbon. In addition, biological growth could also "blind" the carbon and significantly reduce its effectiveness to remove organics.

If the contaminants are inorganics, ion exchange may be applicable. Ion exchange is the exchange of an ion with a high ion-exchange selectivity for an ion with a lower selectivity. Solid phase materials that can exchange an ion in the water for one of its ion may be suitable for incorporation into a reactive wall. Any divalent ion usually will have a higher ion exchange selectivity than will a monovalent ion. Most metals present in ground water are in the divalent or trivalent state and are amenable to ion exchange. However, ion exchange can be expensive and, in the case of activated carbon, it is a physical process which is reversible so the spent material must eventually be removed and managed properly, and naturally occurring constituents such as dissolved iron and bacteriological growth can "blind" ion exchange material reducing its effectiveness.

Biological Degradation

Chapter 3 discusses biological processes in depth so here we simply want to relate potential applications to reactive walls. Biological degradation applications to reactive walls can be accomplished in many different ways. The two main criteria for a biological reactive wall are surface area and oxygen. The bacteria need the proper amount of surface area in order to attach and interact with the plume. If there is too little surface area (i.e., large gravel), an insufficient amount of bacteria will be in the reaction zone. If there is too much surface area (i.e., silt and/clay), the water will not travel past and interact with all of the bacteria.

The oxygen can be supplied in many ways. An air sparging curtain can be installed in natural or geologically altered (to increase permeability) material to supply oxygen to enhance degradation of aerobically biodegradable compounds such as Benzene Toluene Xylene (BTX). Research has been performed (Bianchi-Mosquera et al., 1993) on using solid phase oxygen-release compounds (ORC) to enhance degradation of BTX. ORCs are commonly in the form of concrete briquettes that slowly release oxygen to ground water. Pilot studies have demonstrated their effectiveness in helping to degrade benzene and toluene (at 4 mg/L each) *in situ* for periods of at least 10 weeks. Although promising, the early studies indicated that the bioremediation zone downgradient of the ORC was fairly narrow, so efforts are underway to try to achieve better mixing with contaminated ground water.

Field studies also have been completed based on the work of Devlin and Barker (1993) on use of a reactor wall with a permeability substantially greater than the natural material to circulate nutrients via injection and withdrawal wells within the wall. The system would be operated in a pulsed mode with nutrient-amended water carried downstream of the wall under natural ground water gradients. These results showed early promise, and additional research is underway on its application.

This area is probably the second most active design area behind iron dehalogenation. Once again, the reader will have to review published papers to watch the advances in this technology.

DESIGN CONSIDERATIONS

There are a number of factors that must be carefully considered in the design of reactive walls, particularly walls incorporated into funnel-and-gate configurations. Of utmost importance is that the system manages the entire plume. Since ground water flow directions can vary, it is extremely important that these fluctuations be accounted for. Secondly, the residence time in the reactive wall must be sufficient to achieve the desired reduction in contaminant concentration. Lastly, the ratio of cutoff walls to reactive walls in a funnel-and-gate system should be optimized to minimize costs. To evaluate these considerations, flow and transport modeling is almost always needed to optimize the design.

One of the most critical factors in reactive wall design is the relationship between residence time of contaminated ground water in the reactor and the rate of contaminant degradation that occurs within the reactor. On the one hand, the designer is trying to maximize discharge through the gate to maximize the width of the capture zone, while, on the other hand, the designer wants to maximize residence time in the reactor to ensure adequate attenuation. Since these are inversely related, they must be properly balanced. Again, flow and transport modeling can be very useful in evaluating these relationships.

As one might expect, the discharge through the gate increases with increased hydraulic conductivity of the gate; however, relatively little increases occur with conductivities greater than ten times the aquifer. This is important because high hydraulic-conductivity porous media usually have large grain sizes and low surface area-to-mass ratios. Since reaction rates in most *in situ* media are proportional to surface area, high hydraulic conductivities would result in slower reaction rates.

The rate of contaminant degradation that occurs within the reactive wall is an important factor. The actual retention time in the reactive wall is a straightforward calculation and is obtained by dividing the pore volume of the reactive wall by the discharge through it. For degradation processes that are first order reactions, the retention time necessary is given by the formula:

$$N_{1/2} = [\ln (C_{eff}/C_{inf})]/\ln (1/2)$$

where

$N_{1/2}$ is the number of half lives required.
C_{eff} is the concentration of the desired effluent.
C_{inf} is the concentration of the influent.

To put this in perspective, the chlorinated compound TCE has a half life of 13.6 h so to reduce a plume from 1000 ppb to 5 ppb would require 7.6 half lives or a residence time of 4.3 days. Of course this is a theoretical calculation, and field results have yet to achieve this. Nevertheless, progress is being made in understanding the kinetics involved within the reactive wall, and as these are better understood the likelihood of success will increase and factors needed for design will become available.

A key design consideration is to increase attenuation in the reactive wall, and there are two ways to achieve this. The first way is to increase the reaction rates. This would most commonly be achieved by increasing the surface area of the reactive material — which goes back to the gate (reactor media) not needing to be more than 10 times more permeable than the aquifer — or by using a higher proportion of reactive material in the reactive porous media. The second way is to increase the retention time within the reactor by decreasing the discharge through the gate (more gates) or by increasing the pore volume of the gate by making it longer or wider.

CASE STUDY: REACTIVE WALL DESIGN

Background

Pentachlorophenol (PCP) and tetrachlorophenol (TCP) were detected in onsite ground water samples at a former wood-treating facility in the western U.S. The ground water system consists of a shallow aquifer containing a heterogeneous mixture of marine deposits and artificial fill which is underlain by low permeability siltstones and mudstone. The shallow aquifer ranges in thickness from 10 to 20 ft and averages 15 ft; its saturated thickness averages 7 ft. Based on the results of the site investigation, it was determined that impacted ground water had the potential to move off-site and adversely affect downstream domestic users of the resource (Figure 8). Consequently, remedial action goals were developed to protect the quality and quantity of the resource.

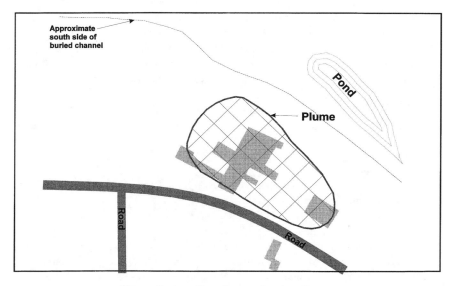

Figure 8   Location of ground water plume.

A number of remedial alternatives were evaluated including pump and treat, containment, and *in situ* treatment. A reactive wall incorporating activated carbon was deemed the most attractive alternative and additional analyses were conducted to determine constraints on below-ground application of this technology and to evaluate its suitability for incorporation into a funnel-and-gate system with the funnel portion comprising slurry wall technology. These studies focused on the following: (1) the potential for underflow beneath the barrier wall, (2) the spatial relationship between gates and funnels, (3) mass loadings on the gates, and (4) interferences with gate performance.

Funnel and Gate Modeling Study

Since reactive walls are more expensive than impermeable barriers, groundwater flow modeling was conducted to optimize the design by minimizing the number and length of gates while still accommodating flow from the entire plume and providing adequate residence time within the reactive gate. Groundwater modeling and particle tracking were used to delineate flow paths through and around the funnel-and-gate system to demonstrate the efficacy of a given configuration. A conceptual model was developed based on varying hydrogeologic conditions and groundwater flow gradients found across the site. The background transmissivity for the site was modeled at 210 ft²/day with two imbedded permeability inhomogeneities of 28 ft²/day and 3.5 ft²/day.

Groundwater streamlines were generated for several combinations of gate lengths and spacings. After several alternatives were evaluated, an L-shaped configuration was selected which contains four gates, two of them installed adjacent to the buried stream channel where flow rates are expected to be higher (Figure 9). The modeling results showed that flow from a fairly broad area can readily be focused through a small number of small gates without causing groundwater seeps to occur at the surface and with minimal disruption of groundwater flow patterns.

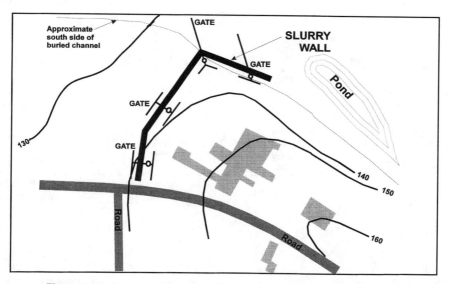

Figure 9 Water-level contours and location of Funnel-and-Gate® system.

The modeling results were also used to evaluate the volume of water expected to be captured for treatment. Based on the area encompassed within the flow lines and the hydraulic properties within this region, a flux of 22 gpm was calculated, or about 5.5 gpm per gate.

Gradient Control

Because of the nonuniform distribution of hydraulic transmissivities across the site, water may be impeded in its lateral movement toward and away from the gates. In order to minimize this effect, gravel-filled collection and distribution galleries will be installed at each gate to collect water from the upgradient side of the gate, guide it through the gate, and then redistribute it uniformly after treatment (Figures 9 and 10). The collection and distribution galleries will be downcut into the aquifer to expose a large cross-sectional area to groundwater flow. This will minimize the pressure required to move water from the aquifer to the carbon treatment gate. Installation of these collection and distribution galleries will help ensure that the pressure drops across the wall will minimally affect the natural groundwater gradients and flow patterns.

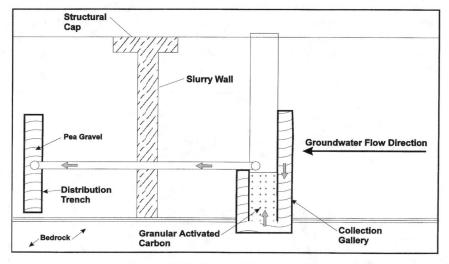

Figure 10   Profile view of treatment gate.

Underflow of Barrier

The potential for flow under the barrier wall also was evaluated by using data collected during aquifer tests conducted at the site along with results from the modeling study. Aquifer tests of the upper marine terrace sediment yielded a high estimated transmissivity of 210 ft^2/day and the transmissivity of the lower Franciscan formation was estimated from slug tests conducted on wells completed solely within the underlying Franciscan formation. A comparison of these two values indicates that the hydraulic conductivity of the lower Franciscan formation is approximately 1/1000 that of the overlying marine terrace sediment. The flows that are expected to travel through the treatment gates were compared with the flows that may flow under the barrier. The solution below for the flow beneath the barrier is similar to that for estimating flow under a dam, with a simplifying assumption that flow lines in the Franciscan are concentric semicircles when viewed in section from one end of the wall.

k = 0.004 (hydraulic conductivity of Franciscan, in ft/day).
i = 0.05 (hydraulic gradient).
T = 210 (transmissivity of marine terrace, in ft²/day).
w = 2 (width of wall, in ft).
h = 2.0 (average head difference across wall, in ft).
x = 400 (distance upgradient from wall, in ft, through which water permeates the Franciscan).

Let q be the flow through a 1-ft-thick slice under the slurry wall emanating from the shallow aquifer between the wall and a point x ft upgradient of the wall.

$$q = \frac{2kix}{\pi} + \frac{k(h - wi)}{\pi} \ln \left(\frac{\pi x}{2} + \pi x \right)$$

$$q = 0.07 \ \text{ft}^3/\text{day}$$

Let Q be the flow through a 1-ft thick slice of the marine terrace deposits.

$$Q = TI$$

For the parameters as defined above, the flow rate is

$$Q = 10.5 \ \text{ft}^3/\text{day}$$

Since the K value for the Franciscan is a horizontal conductivity, the actual flow (q) under the wall will be less than the computed value, because the effective conductivity is a weighted average of the horizontal and vertical Ks, and the vertical conductivity is likely much less than the horizontal conductivity. The maximum flow under the wall represents approximately 0.7% of the total flow through the system.

Gate Design

The gate design incorporates several factors including contaminant concentration, flow rate, and time between carbon changeout. Using an estimated flow rate of 5.5 gpm, a carbon changeout of 3 yrs, a concentration of 200 ppb, and a 1% adsorptive capacity, approximately 48 ft³ of carbon will be required for each gate.

To minimize flow velocity and maximum residence time, each gate was designed using 4-ft diameter corrugated metal pipe filled with 4 ft of activated carbon. This configuration will allow water to move easily through the gates with a pressure loss of only about 2 inH$_2$O.

Ground water samples will be collected from the monitoring points located before each of the treatment gates to measure influent concentrations entering the gate. A sample will then be collected from the monitoring point within the treatment gate to verify that the treatment gate is being effective in removing these compounds prior to the water exiting the gate. A treatment buffer zone exists downstream of this midgate monitoring point to ensure that impacted water does not exit the gate prior to full removal of these compounds. Upon detection of a compound of concern at a concentration above water-quality objectives at the midgate measuring point, the carbon will be removed and replaced. It is expected that biological activity will also be a part of the removal mechanism for the contaminants.

REFERENCES

Bianchi-Mosquera, G.C., Allen-King, R.M., and Mackay, D.M. "Enhanced Degradation of Dissolved Benzene and Toluene Using a Solid Oxygen-Releasing Compound." *Ground Water Monitor. Remed.*

Blowes, D.W. and Ptacek, C.J. "Geochemical Remediation of Ground Water by Permeable Reactive Walls: Removal of Chromate by Reaction with Iron-Bearing Solids." Proc. Subsurface Restoration Conference, June 21–24, 1992, Dallas, TX; Available from Rice University, Dept. of Environmental Science and Engineering, Houston, TX.

Devlin, J.F. and Barker, J.F. "A Semi-Passive Injection System for *In Situ* Bioremediation." Proc. National Conference on Hydraulic Engineering and International Symposium on Engineering Hydrology, July 25–30, 1993, San Francisco, CA, American Society of Civil Engineers, Hydraulics Division.

Gillham, R.W. and Burris, D.R. "Recent Development in Permeable *In Situ* Treatment Walls For Remediation of Contaminated Groundwater."

Gillham, R.W., O'Hannesin, S.F., and Orth, W.S. "Metal Enhanced Abiotic Degradation of Halogenated Aliphatics: Laboratory Tests and Field Trials." Proc. HazMat Central Conference, Chicago, IL, March 9–11, 1993.

Puls, R.W., Powell, R.M., and Paul, C.J. "In Situ Remediation of Ground Water Contaminated with Chromate and Chlorinated Solvents Using Zero-Valent Iron: A Field Study." Proc. 209th American Chemical Society National Meeting, Anaheim, CA, April 2–7, 1995.

Schwartz, F.W. and Xu, Y. "Modeling the Behavior of a Reactive Barrier System for Lead." Proc. Modern Trends in Hydrogeology, 1992 Conference of the Canadian National Chapter, International Association of Hydrogeologists, Hamilton, Ontario, 1992.

Starr, R.C. and Cherry, J.A. "*In Situ* Remediation of Contaminated Ground Water: The Funnel-and-Gate System," *Ground Water: A Journal,* Association of Groundwater Scientists and Engineers.

10 MISCELLANEOUS *IN SITU* TREATMENT TECHNOLOGIES

Frank J. Johns II and Evan K. Nyer

INTRODUCTION

This chapter focuses on other *in situ* treatment technologies for soil and ground water that have not been previously discussed in this book. Many of these technologies are emerging technologies for which there is little or no full-scale experience. Therefore, little can be presented here other than a general description, advantages and disadvantages, information on vendors, basic design considerations including potential areas of application, and some rough estimates of the costs.

There are two distinct categories of technologies presented in this chapter for both soil and groundwater treatment. The first category of technologies includes those that result in the removal of contaminants from the affected media. In the case of soil, these technologies are typically coupled with a demonstrated *in situ* removal technology (e.g., soil vapor extraction) and are used to enhance the performance of the carrier, water or air. The second category of technologies are true *in situ* technologies, but the treatment results in the contaminants being stabilized or "locked" into the soil matrix so that there is no longer a threat of migration to or through the ground water. Therefore, the treatment does not reduce the amount of the contaminant, only the mobility; in some cases, the toxicity is reduced.

CONTAMINANT REMOVAL TECHNOLOGIES

The following miscellaneous *in situ* treatment technologies for soil all involve enhancing the mobility of the contaminant so that it can either be more readily removed by soil vapor extraction or transported into the ground water for subsequent treatment with the ground water. The processes that involve increasing the

0-87371-995-6/96/$0.00+$.50
© 1996 by CRC Press, Inc.

volatility and therefore the vapor phase mobility of the contaminant use the application of heat to the soil (steam flushing, hot air technology, and radio frequency heating). Not only is the application of heat effective in increasing the volatility (see Table 1 for the relation of temperature to Henry's Law constant), but an increase in temperature also decreases the adsorption of the contaminants onto the soil particles and increases the rate of diffusion through the soil (Smith and Hinchee 1992). Other processes associated with the transport of contaminants in the vapor phase for subsequent removal by vapor extraction are electro-acoustic soil remediation and electro-osmosis. Chemically-enhanced flushing is a miscellaneous treatment technology that promotes the movement of soil contaminants into the ground water for subsequent capture and treatment.

Table 1   Relationship of Henry's Law
          Constant to Temperature

Compound	10°C	20°C	40°C
TCE	328	544	1370
Benzene	133	230	619
1,2-Dichloroethane	30	51	134
Methylene Chloride	53	89	226

Steam Flushing

Steam, a well understood and available source of heat, is injected into the contaminated soil to increase the temperature in the contaminated zone, thereby increasing the volatility of the contaminants and also increasing the mobility of the contaminants. The soil is heated as the steam condenses releasing the latent heat. As the steam progresses farther into the soil, it cools enough to become water. Therefore, it is necessary to form a steam front that uniformly permeates the contaminated area and moves from the injection point to the extraction point.

Technology Description

There are two different technologies for in situ steam stripping that have been applied to contaminated soils. The first uses a well or series of wells for steam injection and relies on vacuum extracted from a well or series of wells to draw the steam through the contaminated zone (EPA, 1994a). This technology clearly shows that steam is just a means of enhancing a conventional soil vapor extraction system. The second technology is a licensed technology that uses a rotating, hollow shaft with drill bits to inject the steam into the contaminated zone. (EPA, 1994d; dePercin, 1991). A vacuum is applied at the surface to pull the steam and volatilized contaminants from the soil. A third steam stripping technology that is applicable to ground water and saturated soils is presented later in this chapter.

Design Considerations

The first technology uses standard soil remediation equipment, with the exception of the steam generation. Figure 1 illustrates a steam flushing system (EPA, 1994b). Steam is generated at the surface using a steam generator and water. A water supply, water conditioning system (depending on the quality of the water supply), and fuel source are required for steam generation. The steam is then injected into a well or series of wells. Extraction wells are used to withdraw the steam and volatilized contaminants from the soil and are similar to vapor extraction wells. In saturated soils, the extraction wells are also used to pump ground water. As with soil vapor extraction (SVE), the spacing of the inlet and extraction wells depends on the permeability of the soil formation. The vapor extraction system is similar to a soil vapor extraction system, except that a condenser is required ahead of the moisture separator, or knockout drum. While all SVE systems require moisture separators, the steam injection significantly increases the amount of moisture that must be handled. Steam can also be used to enhance gas and liquid movement at the same time (Figure 2).

The hollow shaft technology was developed by NOVATERRA, Inc., and is implemented using a transportable unit called the Detoxifier™. NOVATERRA should be contacted directly for site specific design considerations. The Detoxifier™ unit (Figure 3) unit consists of both the process tower and the process train. The process tower includes two 5-ft diameter, counter-rotating, hollow-stem drills capable of drilling to a depth of 27 ft and an oil-fired boiler for steam generation. Steam is delivered to the top of the drill, is conveyed down through the outer pipe of the hollow-stem drills, and is injected into the soil through the cutting blades. A shroud on the surface collects the steam and volatilized organic compounds. The process train includes a condenser, a distillation column to separate the contaminants from the condensate, and granular activated carbon (GAC) to treat the distillate. Steam generated on the process tower is used to regenerate the GAC. The unit treats an area of approximately 7 ft by 4 ft to the desired depth (up to 27 ft) on a batch basis.

Either steam flushing technology can effectively remove volatile organic compounds (VOCs) and semivolatile organic carbon (SVOCs) from vadose zone soils. NOVATERRA reports that their system removes VOCs, but only reduces the concentration of SVOCs. The technologies also are capable of treating saturated soils. Because of the drilling unit, the Detoxifier™ is limited to areas where there are no underground obstructions greater than 12 in. in diameter, there is adequate overhead clearance, and the ground slope is acceptable for the rig. The Detoxifier™ can be modified to inject oxidants and/or nutrients to promote bioremediation or to inject stabilization agents for *in situ* stabilization (see below).

Costs

The cost for the steam flushing technology is similar to the cost of a soil vapor extraction system with air inlet wells plus the cost for the steam generator,

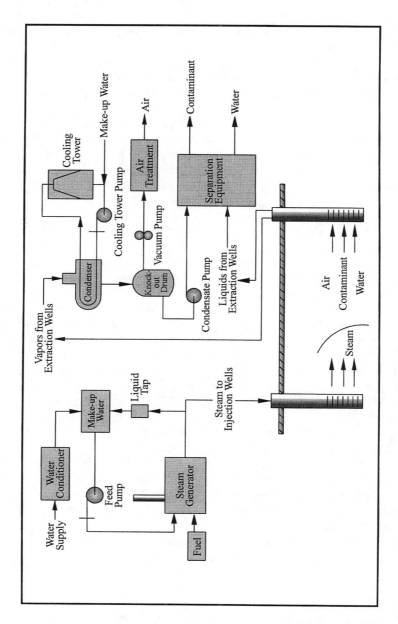

Figure 1   *In situ* steam-enhanced extraction process. (Courtesy of Udell Technologies, Ltd.)

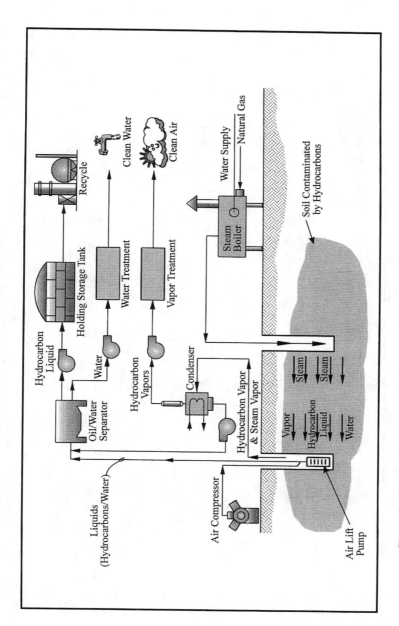

Figure 2 Steam-enhanced recovery process. (Courtesy of Hughes Environmental Systems, Inc.)

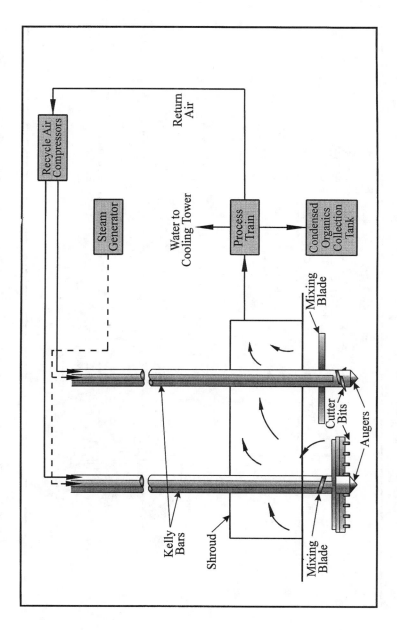

Figure 3   Detoxifier™ process schematic.

steam piping, and the condenser prior to the typical vapor treatment system. The cost for a steam generator and a condenser will depend on the size of the project. Cost information is available from NOVATERRA, Inc., for its technology for specific sites.

Advantages and Disadvantages

Steam flushing has been used in the petroleum industry to enhance the recovery of oil. The experience gained from application of this technique for cleanup of crude oil can be applied to the remedial area. Limitations on the use of steam flushing are similar to the limitations for conventional soil vapor extraction, e.g., inhomogeneity in the formation and impermeable layers can impact the flow of steam. The primary advantage over conventional soil vapor extraction is the ability to remove SVOCs as well as VOCs. The primary disadvantages are the increased capital cost for the steam generation equipment and the increased operating cost for the energy. In addition, there may be a concern with the mobilized contaminants moving to the ground water with the condensed steam. Also, the elevated temperatures may sterilize the soil, thereby requiring that the soil be inoculated before proceeding with bioremediation following the removal of the main mass of the contamination.

Hot Air Flushing

Hot-air flushing applies the same basic principal of steam flushing — increase the soil temperature to enhance soil vapor extraction — but it uses hot air as the heat source instead of steam. The advantage of hot air over steam is that a condensed water is not created, which eliminates the need for water treatment at the surface or the risk of contaminated condensed water entering the aquifer. The disadvantage is related to the thermal properties of air and water. Air has a significantly lower heat capacity than water, so a very high air temperature is required to provide enough heat to uniformly heat the soil in the contaminated zone. Steam can be used for heating in saturated soils, whereas hot air is ineffective in saturated soils because of the large difference between the heat capacity of the air and the water in the soil.

Technology Description

A well or series of wells is drilled to a depth below the zone of contamination in the soil. The wells are perforated at the bottom and cemented above the perforations to prevent hot air from short circuiting through the bore hole. Hot compressed air (250 to $1200°F$ and 5 to 22 psig) is injected into the well(s). As the hot air moves upward through the soil, the soil moisture evaporates and the VOCs and SVOCs volatilize and move upward with the hot air. The hot air and removed organic contaminants are collected in some form of a soil vapor extraction system.

Design Consideratons

The HRUBOUT® process is a patented hot-air injection process (Figure 4) (EPA, 1994e). It employs soil gas collection channels at the ground surface for collection of the hot air and organic contaminants. The process is reportedly capable of removing both halogenated and nonhalogenated organic compounds including gasoline, diesel, jet fuel, crude oil, lubricants, solvents, and creosotes. The HRUBOUT® process employs an air blower, an adiabatic burner for heating the injection air, and an incinerator for off-gas treatment.

The primary design considerations are well spacing, blower capacity, and air temperature. The well spacing depends on the soil permeability and the depth of contamination. The blower capacity depends on the number (air flow) and depth (pressure) of the wells. The air temperature depends on the soil mass, moisture content, and the type of contaminant (VOCs or SVOCs).

Costs

Cost information is available from Hrubetz Environmental Services, Inc., Dallas, TX, for specific sites.

Advantages and Disadvantages

Hot air flushing has the advantage over steam flushing of not requiring a water supply. The burner for heating the air is relatively easy to operate. The primary disadvantage, the limited heat capacity of air, was discussed earlier. Aside from enhancing the removal of less volatile constituents, another advantage of hot air flushing over conventional soil vapor extraction is the enhanced promotion of biodegradation at elevated soil temperatures.

Resistance Heating

Resistance heating is another method used to heat soil *in situ* (Smith and Hinchee, 1992). As the soil is heated, the volatility of VOCs and SVOCs increases making them more susceptible to removal by soil vapor extraction. Resistance heating is used in the *in situ* vitrification process discussed later in this chapter.

Technology Description

Electrodes are inserted in the ground and an electrical current is applied through the electrodes using the soil matrix as a conductor. Water in the soil matrix is the major conducting pathway. The resistance to the flow of current in the soil matrix causes heat to be generated. As the soil is heated, the soil moisture evaporates and the organic compounds volatilize. The water vapor and organic contaminants are removed from the ground by some form of soil vapor extraction. A variation on resistance heating to enhance vapor extraction includes the added step of increasing the voltage to either pyrolize or electrically enhance the oxidation of nonvolatile

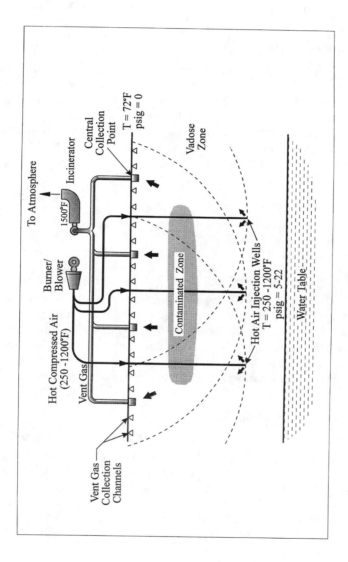

Figure 4 HRUBOUT® process.

organic compounds, creating more volatile breakdown products that can be removed through vapor extraction.

There are various types of vapor extraction systems that can be applied with resistance heating. For lower temperature-resistance heating applications, a conventional well point system can be used. For high temperature *in situ* vitrification applications, a vacuum hood is placed over the soil mass being heated.

Design Considerations

Water in the soil matrix is the major conducting pathway; therefore, as the soil moisture decreases, the conductivity of the soil is reduced. This can lead to uneven heating of the soil matrix. This problem needs to be addressed through the proper design and placement of the electrodes.

Bench-scale tests have shown that resistance heating technology is applicable to soil types ranging from sands to clays. Removals rates up to 99% for SVOCs and nonvolatile compounds have been seen.

Costs

Only pilot-scale testing has been performed in the field. As yet, no costs have been published for this technology.

Advantages and Disadvantages

The electrodes are small diameter rods or tubes that are easily inserted into the ground; therefore, it is a much easier setup for soil heating when compared to steam and hot-air flushing. In addition, the installation and operation of an electric power supply is much simpler than that of either a steam generator or a burner to heat the air. But, there is no full-scale experience with resistance heating for enhancing soil vapor extraction, which is a significant disadvantage.

Radio Frequency Heating

Radio frequency (RF) heating is an *in situ* process used to heat the soil matrix to temperatures greater than $100°C$ to enhance the volatility of organic compounds for removal by soil vapor extraction (Smith and Hinchee, 1992). The primary mechanism of contaminant removal is vaporization, but, depending on the temperature, organic compounds can be decomposed or pyrolized to form more volatile compounds and, depending on the temperature and soil moisture, steam distillation can occur to remove the compounds from the soil particles and move them toward the surface of the ground.

Technology Description

Radio frequency heating uses an array of excitor electrodes placed either vertically in the ground or horizontally along the ground surface. A radio frequency transmitter is used as the power source and is connected to these electrodes. The soil is heated through two mechanisms: resistance heating and dielectric heating. Electrons moving though the soil mass cause resistance heating to occur. Dielectric heating is created by the magnetic distortion of the atomic or molecular structure in response to the applied electric field. The distortion is created in polar compounds, primarily water, as the dipoles align.

The volatilized organic contaminants are removed from the soil using conventional soil vapor extraction. When the electrodes are placed on the ground surface, the ground surface is covered with some sort of synthetic membrane to provide a surface seal. Vapors are collected under this membrane and air is drawn across the ground surface to a vapor collection and treatment system using an air blower.

Design Considerations

Radio frequency heating is applicable for the removal of higher boiling point organic contaminants in unsaturated soils. During treatment, the soil is heated to temperatures ranging from 100 to 400°C. In studies conducted at a temperature of approximately 150°C, the VOCs were removed at a rate of between 95 and 99% and SVOCs were reduced from 90 to 95%.

Radio frequency heating is controlled by the RF frequency and the field strength, which is limited by the power of the transmitter. The selected frequency depends on the soil properties and the depth of contamination. The soil properties affect the response time of the dielectric movement of the molecules. The frequency needs to be tuned to maximize the movement. Moisture content is the primary soil property of concern. As heating proceeds, moisture is driven off and the dielectric properties of the soil change. To account for this, the frequency must be changed, requiring fairly sophisticated operational control. The frequencies used range from 45 Hz to 10 GHz, with a primary range of 6.8 MHz to 2.5 GHz used on most remediation sites.

Costs

Costs have been calculated by IIT Research Institute, Chicago, IL, for treating a 1-acre site to a depth of 20 ft. The soil temperature was increased to approximately 300°C. The length of time for the soil to be heated depended on the power of the transmitter. Using 2 MW of power, the soil was heated to 300°C in approximately 250 to 400 days. Using 10 MW, the time was decreased to between 60 and 90 days. The cost ranged from $50 to $100 per ton depending on the moisture content of the soil (Harsh et al., 1984).

Advantages and Disadvantages

The primary advantage of RF heating is that it allows for the removal of higher boiling point compounds than does steam flushing. The disadvantages include the high equipment costs, the intensive operator control required, and the possibility that the high temperatures may sterilize the soil, inhibiting future biodegradation. In addition, the technology has limited field experience.

Chemically-Enhanced Flushing

Flushing is a method to remove and capture contaminants in the soil. It is also referred to as soil flushing. Soil flushing uses water, chemically-enhanced water, or gaseous mixtures to carry the contaminants to a point where they can be collected. The process accelerates a number of subsurface contaminant transport mechanisms found in standard processes that use water and air as carriers. The application of the process requires a sound understanding of the manner in which target compounds are bound to the soils and hydrogeologic transport.

Technology Description

As we discussed in Chapter 1, water and air normally are used as carriers to remove contaminants from the ground. Soil flushing is the enhanced use of these carriers to remove the contaminants. In its simplest form, soil flushing uses large amounts of water to remove the contaminants from the soil. Techniques such as water flooding can be used to significantly increase the amount of water passing through the soil. Water also can be chemically enhanced to carry more of the contaminant (Figure 5).

Each contaminant has a natural solubility in water. Organic compounds can be carried in water to the limit of their solubility. Chemicals (surfactants, for example) can be added to the water to increase the amount of contaminant that can be carried by the water. The chemicals can also change the adsorption/desorption, acid/base reactions, and biodegradation of the contaminant in the soil. The combination of these mechanisms allows the contaminants to be removed at a much faster rate than the carrier can perform by itself.

Design Considerations

There are two main design considerations with soil flushing. First, the geologic conditions at the site must accommodate the use of the carrier to reach and remove all of the contaminant. This requires that the soil be homogeneous and permeable. Macrogeological conditions such as sand and clay lenses or manmade conditions such as pipes and utilities will create a path for the carrier that does not interact with all of the contaminated soils. Also, the microgeologic limitations that were discussed in Chapter 1 also will limit the effectiveness of the process. The technology should be limited to sands and certain silty sands for the process to be effective.

Second, the contaminant must be amenable to the enhancement. Contaminants that are very insoluble or tightly bound to the soil will not be significantly affected by the chemical enhancement. The process is being considered for applications involving petroleum hydrocarbons, chlorinated hydrocarbons, metals, salts, pesticides, herbicides, and radioisotopes.

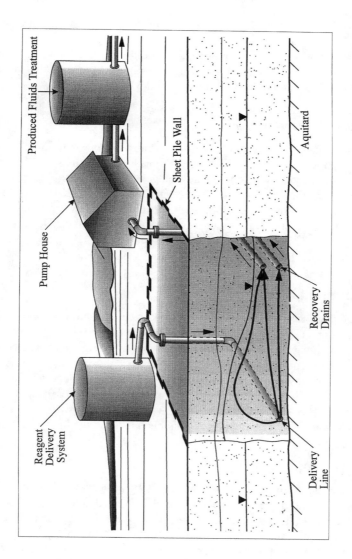

Figure 5 Soil flushing.

Successful applications of this technology have been limited to bench-scale tests. Field applications have not been successful due to reduced permeability by plugging and biofouling. Also, most of the carrier seems to end up moving through a few major flow paths. Further work needs to be performed to better understand the flow limitation in field applications.

Costs

Capital and operating costs are similar to traditional pump-and-treat systems, with the added cost of chemical application. Estimated costs are between $60 and $130/m^3 or $80 to $165/yd^3 (Anderson and Mann, 1993).

Advantages and Disadvantages

The main theoretical advantage of chemically-enhanced flushing is that it allows the removal of the contaminants without disturbing the soil. This should provide environmental and economic advantages over traditional processes. The enhancements should also increase the rate in which a compound can be cleaned from a site, allowing the remediation to occur in significantly less time.

The main disadvantage is that the process still relies on the carrier to remove the contaminant. This has caused severe problems when the process has been taken to the field.

Electrochemical Remediation

Electrochemical remediation (ECR), also known as electrokinetics or electro-osmosis, is a process in which an electrical field is created in the soil matrix by applying low-voltage, direct-current electrical power to electrodes placed in the soil (Figure 6) (Trombley, 1994). When the electric current flows between the anodes and cathodes, heavy metals, radionuclides, and soluble organics migrate toward the electrodes. The contaminants are deposited near the electrodes and can be removed by conventional methods, e.g., excavation, or by circulating a fluid at the electrode and removing the contaminants from the fluid (EPA, 1994f).

Technology Description

The electrodes are placed in the soil in either a vertical or horizontal array. When direct-current power is applied to the electrodes, an electric field develops between the anodes and cathodes. Studies have shown that moisture in the soil forms an acid front at the anode and migrates toward the cathode. The two primary transport mechanisms for the contaminants and the soil moisture containing dissolved contaminants are electromigration and electro-osmosis.

Electromigration is simply the term used for the movement of cations and anions in the electric field in the direction of one or the other electrodes. Since the electromigration mechanism acts on charged particles, it primarily affects ionized inorganic compounds that are also highly soluble. These compounds

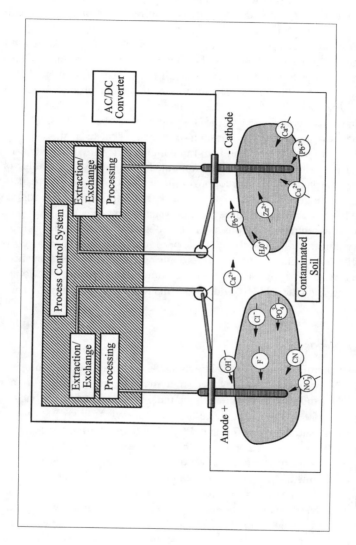

Figure 6 Electrokinetic remediation process. (Courtesy of Electrokinetics, Inc., Baton Rouge, LA.)

include cations, such as the alkali metals, and anions, such as chlorides, nitrates, and phosphates.

Electro-osmosis is a transport mechanism whereby water moves through charged soil. This mechanism is most effective in fine-grained soil. As the water (acid front) moves from the anode to the cathode, contaminants are carried with it. This is the primary removal mechanism for soluble organic contaminants. The treated soil matrix is acidified, but returns to equilibrium rapidly through diffusion of other anions and cations once the electric source is removed.

Several modifications to the basic ECR process are in various stages of development. Monsanto, Dupont, and General Electric entered into a Cooperative Research and Development Agreement with the U.S. Environmental Protection Agency (EPA) in 1983 to develop an *in situ* treatment process based on ECR (*Environ. Progress,* 1994). The Departments of Energy (DOE) and Defense (DOD) have also contributed to the development of the process. The modified ECR process is known as the "Lasagna process" because of the layered nature of the treatment mechanism. Solid or granular electrodes are installed horizontally or vertically on either side of the contaminated soil. Highly permeable zones are established through the construction of horizontal layers or vertical trenches between the anode and cathode. Once current is applied and the contaminants migrate, they move through these porous zones where they can be treated using more standard treatment methods such as biodegradation, catalytic dechlorination (for Polychlorinated Biphenyls (PCBs)), or granular activated carbon adsorption.

Batelle Memorial Institute is developing a patented electroacoustic process that enhances ECR by using an acoustic source between the electrodes (EPA, 1994g). The acoustic field is added to enhance the dewatering of sludge-like wastes.

Design Considerations

Electrochemical remediation is applicable to highly soluble inorganic contaminants; heavy metals, such as lead, mercury, cadmium, chromium, and zinc; radionuclides; and soluble organic compounds. Removal rates have been reported in the range of 50 to 99%. Specific studies have shown heavy metals and uranium removed at a rate between 75 and 95% when the initial concentration was approximately 2000 ppb. Phenol was shown to be removed at a rate between 85 and 95% with a starting concentration of approximately 500 ppb.

Electrochemical remediation is reported to work with both saturated and unsaturated soils. The electro-osmosis mechanism is most effective in moist, fine-grained soils. Many of the studies conducted to date have been conducted on moist, nearly saturated, soils. Large metallic objects such as pipes, fence posts, drums, or other debris can reduce the efficiency of the process.

The type of electrode used is another design consideration. Metallic vs. carbon/graphite and solid vs. granular are the options. Metallic electrodes may dissolve from electrolysis and introduce the corrosion products (e.g., oxidized iron) as a residue into the soil matrix. If a circulating fluid is used at the electrodes, the designer needs to be aware of the chemical properties and the precipitation chemistry of the contaminants to maintain the proper pH of the circulating fluid. Acidification of the soil should not be a long-term problem since it has been shown that equilibrium is restored through diffusion. But, the potential for migration of contaminants resulting from the creation of the acid front needs to be considered.

Costs

Reported field-scale costs range from $90 to $130/ton of treated soil for organic- and chromate-contaminated soil, respectively. A pilot-scale study was conducted on lead contaminated clay and reported electricity costs at $15/ton. Because this is an emerging technology, only limited full-scale costs are available.

Advantages and Disadvantages

Electrochemical remediation has an advantage over conventional pump-and-treat technologies in fine-grained soils. One of the target areas for the technology is the low permeability, silt- and clay-type soils. It has been shown to be effective in causing the contaminants to migrate in these low permeability situations where pumping is slow and difficult. There is a similar advantage realized over conventional soil vapor extraction in the tighter formations. Another advantage over soil vapor extraction and certain other *in situ* technologies is that ECR is capable of removing both inorganic and organic contaminants. ECR has limited full-scale applications. The technology is still emerging, so other advantages and disadvantages will become apparent over time.

In-Well Air Stripping

In-well air stripping uses air to remove the volatile contaminants from the ground water while present in the well. The process is a combination of two basic functions. First, a system is set up to circulate the surrounding ground water through the well. Second, the ground water is put into contact with air, with the volatile contaminants transferring from the water to the air. The air is then collected at the top of the well and treated as required before being released to the atmosphere.

Technology Description

In-well air stripping uses the well to circulate the ground water and to treat the ground water as it enters the well. The process normally uses two sets of well screens to accomplish the circulation. The lower screen is used to draw in the water from the surrounding area of the well. The water is then pumped to the top of the water table, using either air or a mechanical pump to move the water. The water is released to the upper screen in the well to re-enter the aquifer. The water movement from the bottom of the well to the release at the top of the well creates a circulation pattern in the surrounding aquifer.

Depending on the design, air is introduced into contact with water either at the bottom or near the top of the well, or a combination of the two. The bubbles from the air introduced at the bottom are similar to above-ground air stripper designs based on diffused air or bubble aeration. Designs with air introduced near the top of the well are similar to above-ground stripper designs based on packed-tower or tray technologies. The volatile contaminants in the water transfer to the air. The size of the bubbles and the length of contact between the air and water control the transfer efficiency. The air and water separate at the top of the well with air continuing on to treatment in above-ground systems before being released.

Figure 7 shows the Unterdruck-Verdampfer-Brunnen (UVB) process, which uses air bubbles to create the pumping action to lift the water in the well or a combination of air and mechanical pumping for water movement. UVB can be based upon water rising or falling due to the addition of the mechanical pump. The UVB process was developed in Europe and has been licensed to Roy F. Weston, Inc., Woodland Hills, CA and IEG Technologies, Charlotte, NC in the United States.

Design Considerations

The two main design considerations with in-well air stripping are the volatile nature of the compounds and the circulation through the surrounding geology. The stripping of the compounds is directly related to the Henry's Law constant that is used with above-ground air stripper designs. It must be remembered that Henry's Law constant is based on volatility and solubility; therefore, in-well air stripping will not work on compounds that are volatile and also soluble in water, for example, acetone. For compounds that normally are considered for treatment in an above-ground air stripper, the in-well air stripping can be used.

There are limitations to the removal efficiency obtained with in-well air strippers. The amount of air that can be released in a well is limited due to the pumping action that it creates in the well. In addition, the length of contact between air and water are limited to the depth and positioning of the contaminated plume. Demonstration results have shown that the UVB process can remove TCE in groundwater by an average of 94% (EPA, 1995). Other in-well air strippers have been limited to 85% removal without the use of the mechanical pump to increase the contact between the air and water.

The second design consideration is the water circulation through the well. The key to in-well air stripping is the circulation of water affecting a wide area surrounding the well. This is required for each well to affect sufficient area in the aquifer. Good circulation with the surrounding area will have all of the normal geological requirements of any water movement through the geologic zone, i.e., no sand or clay lenses. In addition, the aquifer must be able to handle large amounts of recharge water created by the circulation in the well. High mounding of water over the recharge area will make the process relatively inefficient and limit the area of effectiveness of the well. It is estimated that the maximum area of influence from the UVB design is 60 to 80 ft surrounding the well.

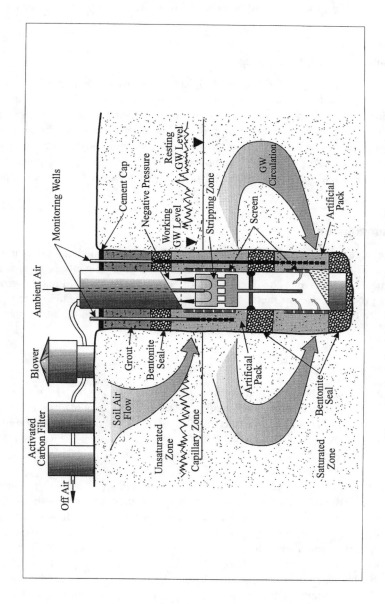

Figure 7 Unterdruck-Verdampfer-Brunnen (UVB) process.

Costs

Most in-well air strippers have only been demonstrated through the pilot stages. Full field application costs are not available for this technology. Specified costs for the UVB process should be obtained from Roy F. Weston, Inc., or IEG Technologies.

Advantages and Disadvantages

The main advantage of the in-well air stripping process is the removal of volatile organics without introduction of air directly into the aquifer. This allows for a more controlled contact between the air and water and an easier capture of the air for subsequent treatment above ground. The disadvantages of the in-well air stripping process are the limited air/water contact and subsequent removal efficiency of the volatile organic compounds. Also, more work is required to determine the actual radius of influence from the groundwater flow through the well.

Phytoremediation

Vegetative remediation is the use of natural plants to remediate a contaminated site. The plants can be used to accumulate the contaminants or to create environments conducive to the degradation of contaminants through natural biochemical processes. Newer methods of phytoremediation have been able to use trees for the dual purpose of direct treatment of contaminants in the aquifer and as pumping mechanisms to control groundwater flow.

Technology Description

All plants have the natural ability to use the root system to acquire nutrients and water from the subsurface. Remediation systems based upon plants were first applied to contaminants that the plants could directly use as nutrients. Ammonia and nitrate plumes both can serve as nutrient sources for all plants. If a plant root system can come in contact with these plumes, then the plant thrives with this added source of nutrient.

Several researchers have shown that the contaminants that can be removed by plants can extend beyond their natural nutrients. Other compounds that have been removed from the soil and ground water include pesticides, heavy metals, and a wide range of organic compounds. Organic constituents can be adsorbed by the vegetation and metabolized or otherwise degraded by contact with the plant root system. Plant roots are extremely valuable to a wide range of microbial populations that thrive in the area immediately surrounding the roots (the rhizosphere). Once in contact with the plant roots, many organic molecules can be subjected to microbial degradation.

In addition, several types of trees have been found to be able to use large amounts of water. Trees that are known for fast growth and high water usage are referred to as "phreatophytes". Common phreatophytes include cottonwood and willow trees. Willow trees have been reported to use and lose over 5000 gallons of water in a single summer day. It has been reported that some phreatophytic trees can have affective root systems up to 100 ft deep into the ground. This water usage allows for a second function of phytoremediation. The trees can be used to contain a ground water by the simple action of their pumping ground water into the atmosphere. At a pumping rate of hundreds of gallons per tree per day pumping rate, a group of trees can have a significant affect on the groundwater flow in tight soils.

The most advanced application of this technology is trademarked TreeMediation™. This technology has been advanced by Applied Natural Sciences, Inc., Columbus, OH. They have been successful in growing tree systems with roots as deep as 30 ft below ground surface. The roots are in contact with the aquifer and use the ground water as their main source of water. TreeMediation™ has been successful in containment of ground water due to the pumping action of the tree and degradation and removal of organic and inorganic compounds (Gatliff, 1994).

Design Considerations

There are two main design considerations for phytoremediation. First, the systems work better in relatively tight soils. In order for the root system and trees to control the aquifer, limited groundwater flow must be present. Most of the proprietary application of this technology is the method of developing the tree root system at the groundwater level as opposed to simply at the surface level. The natural root system in humid areas would be able to satisfy all requirements at or near the surface of the ground. In order for this process to work on contaminants in the ground water, methods must be developed to force the root system down to the ground water.

The second design consideration is the compounds to be treated. Many compounds act as nutrients and enhance the growth of the plants. These compounds are easily treated and contained. The accumulation of heavy metals will depend on the availability of the metals to the plant; we have discussed precipitated metals not being available to the water or biological systems. Nonsoluble metals will be stopped by other processes in an aquifer. Organic compounds can be degraded at the root system, accumulated in the plants, or simply pumped to the atmosphere with the water moving through the trees. The process will depend on the chemistry and properties of the organic compounds.

Costs

Once again, this technology is just now being applied in the field, and exact costs are not available. Gatliff (1994) produced a table summarizing the cost of TreeMediation™ on the 1-acre site with a 20-ft deep aquifer. These costs are summarized in Table 2. As more installations rely on phytoremediation, more accurate economic valuations will become available.

Table 2   TreeMediation™ Costs

TreeMediation™ program design and implementation	$50,000
Monitoring equipment	
Hardware	10,000
Installation	10,000
Replacement	5,000
5-year monitoring	
Travel and meetings	50,000
Data collection	50,000
Annual reports	25,000
Effectiveness assessment — sample collection and analysis	50,000
Total	**$250,000**

Advantages and Disadvantages

The main advantage of vegetative remediation is that it is a natural process that produces no residuals. The process requires no long-term operation and maintenance other than the monitoring analysis that all remediation requires. The process also does not require that the site be stripped and prepared for the use of above-ground equipment. The use of trees is environmentally friendly.

The main disadvantage of TreeMediation™ and other phytoremediation processes is the limitation to near-surface contamination (less than 30 ft), and the application to areas of tight soil. As more work is performed in these areas, researchers may find methods to extend the root systems down to the aquifer and in a high variety of geological conditions. Remediation will probably will be limited to the top portion of the aquifer. Deep plumes will not be remediated with this technology.

CONTAMINANT FIXATION TECHNOLOGIES

Stabilization and Solidification

In situ stabilization and solidification is implemented by applying the same technology to the soil as is done in the *ex situ* process that is commonly applied to soil contaminated by heavy metals or other inorganic compound to immobilize the contaminants in the soil prior to landfilling. Stabilization and solidification is conducted by mixing the soil with additives to produce a cement-like mass. *In situ* stabilization and solidification is conducted by using a device to mix the soil in place with the additives.

Technology Description

In situ stabilization and solidification involves three main components: (1) a means to mix the soil in place; (2) a reagent (or additive) storage, preparation, and feed system; and (3) a means to deliver the reagents to the soil mixing zone.

The means to both mix the soil and deliver the reagents is typically accomplished through the use of hollow-stem auger equipment that is specifically designed for the process. Raw reagent materials are typically stored in silos, and some form of batch plant with mixing tanks is used to prepare the additive slurry. Common additives are cement, lime, cement or lime kiln dust, and fly ash. Proprietary stabilizing agents also are used for certain types of contaminants.

The typical mixing equipment uses multiple (two or four) 3-ft to 12-ft diameter augers. As the augers move down into the contaminated soil, the stabilizing agent is added as the soil is mechanically mixed by the augers. Once the augers reach the lowest extent of contamination, they are removed and the soil is mixed a second time. The size of the area treated in each drilling pass depends on the size and number of augers used. The curing period depends on the stabilizing agent. Once the stabilized matrix is cured, it has unconfined compressive strength, permeability, and leachability similar to *ex situ* stabilized soils.

Design Considerations

In situ stabilization and solidification is applicable to contaminated soils, sediments, and pond sludges. The process is typically used for inorganic compounds but can be applied to organically contaminated soil in some cases. In the latter applications, proprietary stabilizing agents are more commonly used. It is recommended that a simple treatability study be conducted to determine the proper stabilizing agent and the resulting characteristics of the soil matrix. Characteristics that are typically considered include unconsolidated compressive strength, permeability, leachability, and wet-dry and freeze-thaw durability.

In situ stabilization and solidification can be conducted up to depths of 100 ft, depending on the auger mixing equipment. Obstructions in the soil, such as large boulders, pipes, or drums, will make this technology unacceptable. Overhead obstructions need to be considered due to the height of the mixing equipment.

Soil mixing without the addition of stabilizing agents has been used to enhance both soil vapor extraction and *in situ* bioremediation. The soil can be mixed to provide a more homogeneous soil matrix in cases where either the heterogeneity of the soil or clay layers limits the effectiveness of soil vapor extraction. To stimulate the growth of microorganisms in the subsurface, solutions containing oxygen and nutrients can be added to the soil through the hollow stem augers as the soil is mixed.

Companies that can be contacted for site-specific design considerations for *in situ* stabilization and solidification include Geo-Con, Inc. (EPA, 1994h), S.M.W. Seiko, Inc. (EPA, 1994j), and In-Situ Fixation Company (EPA, 1994i).

Costs

A number of *in situ* stabilization and solidification projects have been completed to date and, therefore, there is fairly reliable cost information. Overall treatment costs have been reported as ranging between $100 to $150/yd^3 of soil treated, depending on the size of the mixing equipment. The mixing costs alone have been reported at approximately \$40/yd^3 for larger mixing equipment. The cost of additives can be estimated from information on *ex situ* stabilization projects. The process can treat approximately 100 to 200 yd^3/day.

Advantages and Disadvantages

The primary advantage of *in situ* stabilization and solidification is that it is a proven *ex situ* technology that can be readily applied *in situ*. Because of the extensive *ex situ* experience, there is good data available on the various stabilizing agents. A significant number of *in situ* projects have been completed, and data is available from the commercial companies that have completed those projects. This technology is also potentially applicable to both inorganic and organic contamination. Since the stabilization is performed *in situ*, regulatory concerns related to the excavated soils are eliminated. The technology is cost competitive with the *ex situ* technology, especially for deep soils.

Stabilization and solidification is an immobilization technology and does nothing to reduce the quantity or toxicity of the contaminants in the soil. Therefore, it may be necessary to cap the treated area to ensure that there is no contact with the soil. Another disadvantage relates to the bulking of soil that can occur because of the addition of a stabilizing agent. A bulking factor between 10 and 30% should be used, depending on the soil type and additive. This can be investigated during the treatability testing.

Vitrification

In situ vitrification (ISV) is an *in situ* stabilization process. The technology is used to immobilize inorganic contaminants in the soil. The immobilization is accomplished by heating the soil mass to high temperatures (1500 to 2000°C) to melt the soil. When the soil cools, it forms a glass- and crystalline-vitrified material that encapsulates the inorganic contaminants, thereby making them immobile. Because of the high temperature of the vitrification process, organic contaminants, along with naturally occurring organic compounds, are volatilized and removed through a form of vapor extraction as the temperature increases in the soil mass or are pyrolized at the high temperatures with the gases removed by the vapor extraction process.

Technology Description

The soil is heated in the ISV process using resistance heating similar to the resistance heating process described above under contaminant removal technologies. The temperatures are significantly higher and the process more energy intensive. The power is supplied at the top of the soil to begin the process. Since soil has a low electrical conductivity until it is molten, flaked graphite and glass frit are placed on the soil surface to initiate the melt. Once this melt begins, the electrodes are lowered 1 to 2 in. per hour. The electrode array is lowered until the entire contaminated area is molten. The mass is then allowed to cool into a monolith. A hood is placed over the reactive area to capture the off-gases produced by the process. A negative pressure is maintained under the hood. The gases are treated before being released to the atmosphere. *In situ* vitrification was developed by Batelle's Pacific Northwest Laboratory and is marketed exclusively by Geosafe Corporation, Richland, WA (EPA, 1994k).

Design Considerations

Since ISV is marketed exclusively by Geosafe Corporation, they should be contacted for specific design information related to a site. *In situ* vitrification uses electricity to generate the heat and therefore requires a source of electrical power. Because of the large power requirements (800 to 1000 kWh per ton of soil), a 12.5 to 13.8 kV transmission line is recommended as the power supply.

The limitations to resistance heating were presented above. There are three additional limitations that prohibit the use of ISV: (1) void volume within a melt greater than 150 ft^3; (2) mass of rubble within a melt exceeding 20% by weight; and (3) combustible material within the melt greater than 5 to 10% by weight. The first two limitations are related to the ability to heat the soil to form the melt, while the latter limitation relates to a concern with combustion of the organic material in the soil during the process. *In situ* vitrification is capable of functioning under saturated conditions, but the soil cannot be heated above 100°C until the water boils off. Resistance heating is an expensive means for boiling water, and it may be more cost effective to implement some form of dewatering in the area of the melt prior to ISV.

Most natural soils can be processed using ISV without any modification since they have a good conductivity once they are molten. Therefore, there is only the need to add the flaked graphite and glass frit to start the melting process in most soils. In limited cases, fluxing material may be required within the soil matrix to provide adequate conductivity for melting to proceed. This is more typical of vitrification of sludges and should not be a consideration in most cases. A volume reduction of 20 to 40% is typical and needs to be considered in the design (Amend and Lederman, 1991).

Costs

Cost information is available from Geosafe Corporation for specific sites.

Advantages and Disadvantages

The advantages of ISV over conventional *in situ* stabilization processes include the benefits of removing organic contaminants from the soil where they are present and the durability of the solidified mass. The disadvantage of ISV is the cost of treatment, which is directly related to the high amount of energy used to melt the soil.

In Situ Biochemical Precipitation

In situ biochemical precipitation is the use of biological activity to change the valence state of the metal, thereby changing its properties in the aquifer. One of the best uses of this technology will be for hexavalent chromium (Cr^{+6}). Cr^{+6} is very soluble in water and travels with the ground water. Cr^{+6} is readily reduced to the trivalent form of chromium. Trivalent chrome precipitates at neutral pH and interacts with the soil particles by ion exchange and adsorption. These mechanisms combine to remove Cr^{+3} from the ground water. Therefore, if reducing conditions can be introduced to a Cr^{+6} plume, then the Cr^{+6} will change to Cr^{+3} and natural geochemical processes will remove the heavy metal from the water.

Technology Description

Biogeochemical precipitation uses bacteria to produce the reducing conditions in the aquifer. Easily degradable organics are introduced into the contaminated zone of the aquifer. Natural bacteria degrades the organics and use all of the natural electron acceptors. Once all the oxygen and nitrates are removed from the contaminated zone, then bacteria proceeds to alternate electron acceptors (see Chapter 3 for full description) and the redox potential of the aquifer in that area is reduced. Under reducing conditions, Cr^{+6} is reduced to Cr^{+3}. Cr^{+3} is then removed from the ground water by precipitation and geochemical interaction with the soils. Since the Cr^{+3} is no longer soluble in the water, the chromium plume ceases to exist.

Design Considerations

No full-scale systems of biogeochemical precipitation have been completed. Field trials have been successful at the pilot level. For example, a site in the midwest had a Cr^{+6} plume. Figure 8 shows the extent of the plume, and Figure 9 shows that the plume spread within a sand seam in the aquifer. To facilitate implementation and assessment of the biological *in situ* chromium reduction process, the pilot study required installation of three injection wells and five monitoring wells. The three injection wells were installed within the vacant facility building, and the five monitoring wells were installed along the eastern edge of the facility building, approximately 40 ft downgradient of the injection well (Figure 8). The installed injection and monitoring wells were shallow monitoring wells screened across a 2- to 3-ft thick sand seam at approximate intervals 10 to 15 ft below grade level (Figure 9). Levels of Cr^{+6} measured in samples collected from the injection and monitoring wells prior to initiating the pilot test were as high as 15 mg/L.

To create the reducing conditions required to change the Cr^{+6} to Cr^{+3}, a dilute molasses solution was periodically injected (approximately 10 gal per week per injection well) into the shallow portions of the impacted aquifer via the three injection wells. The molasses, which consist mostly of sucrose, is readily

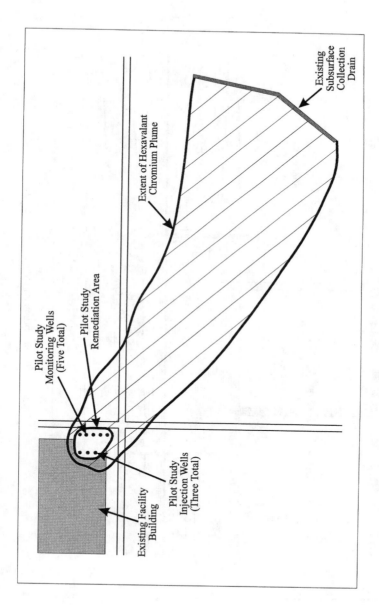

Figure 8 Hexavalent chromium plume.

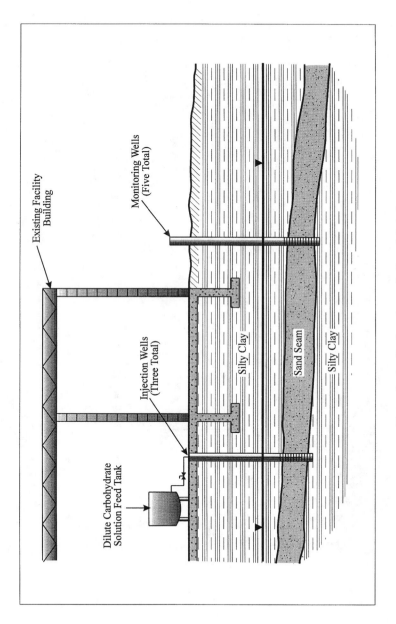

Figure 9   *In situ* chromium reduction pilot study.

degraded by indigenous heterotrophic microorganisms present in the aquifer. During the course of the pilot study, monthly monitoring of the three injection wells and the five monitoring wells was conducted. Within one month of the process initiation, strongly reducing conditions developed in and around the three injection wells, as demonstrated by significant decreases in the levels of dissolved oxygen and oxidation reduction potential (ORP) measured using a down-hole probe. During the first month of operation, concentration of Cr^{+6} in the injection wells decreased from a high of 15 mg/L to below 0.02 mg/L, which is below the ground water cleanup objectives for chromium at this site. At the natural rate of groundwater flow, it is expected that the reducing conditions will reach the monitoring wells within 3 to 6 months. At that time, it would be determined if the biogeochemical reduction can be accomplished over a wide area of the aquifer.

Costs

At the present time, not enough work has been accomplished to complete a cost analysis of this process. However, it is anticipated that a simple injection of sugars will be significantly less costly than the removal of chromium by pump-and-treat systems.

Advantages and Disadvantages

The advantage of the biogeochemical precipitation is the use of simple sugars and bacteria natural to the aquifer to create a reducing environment for the change in valence state of the chrome. Thus, all natural processes are used to facilitate the change for chromium from a soluble state (Cr^{+6}) to an insoluble state (Cr^{+3}). The process should be able to clean an aquifer in significantly less time and for less money than pump-and-treat or other processes. The main disadvantage of this process is that the metal is not actually removed from the aquifer, but only put into a state in which the water carrier will no longer interact with it. It will have to be determined if the sugar solutions can be delivered over a wide range of the aquifer so that the entire plume is affected.

REFERENCES

Abdul, A.S. and Ang, C.C. "In Situ Washing of Polychlorinated Biphenyls and Oils from a Contaminated Field Site: Phase II Pilot Study." *Ground Water,* vol. 32, no. 5, pp. 727–734, 1994.

Amend, L.J. and Lederman, P.B. "Critical Evaluation of PCB Remediation Technologies. Presented at the American Institute of Chemical Engineers 1991 Summer National Meeting, Pittsburgh, August 1991.

Anderson, W.C., ed.; Mann, M.J., Task group leader, *Innovative Site Remediation Technology, Vol. 3: Soil Washing/Soil Flushing,* American Academy of Environmental Engineers, 1993.

de Percin, P.R. "Demonstration of In Situ Steam and Hot-Air Stripping Technology." *J. Air Waste Manage. Assoc.,* vol. 41, no. 6, pp. 873–877, 1991.

EPA, "Demonstration Bulletin: Uuterdruch-Verdampfer-Brunnen Technology (UVB) Vacuum Vaporizing Well." EPA/540/MR-95/500, U.S. Environmental Protection Agency, January 1995.

Dougherty, E.J., McPeters, A.L., Overcash, M.R., and Carbonell, R.G. "Theoretical Analysis of a Method for In Situ Decontamination of Soil Containing 2,3,7,8-Tetrachlorodibenzo-*p*-Dioxin." *Environ. Sci. Technol.,* vol. 27, no. 3, pp. 505–515, 1993.

Gatliff, E.G., Vegetative Remediation Process Offers Advantages Over Traditional Pump-and-Treat Technologies." *Remediation,* Summer 1994.

Harsh, D., Bridges, J.E., and Sretsy, G.C. "Decontamination of Hazardous Waste Substances from Spills and Uncontrolled Waste Sites by Radio Frequency In Situ Heating," in *Proc. Hazardous Material Spills Conference*, Philadelphia, PA, Government Institutes, Inc., April 1984.

"'Lasagna' Process Treats Contaminants." *Environ. Progress,* pp. 69–70, September 1994.

Roy F. Weston, Inc./IEG Technologies, "IVB-Vacuum Vaporizing Well," EPA Superfund Innovative Technology Evaluation Program, EPA/540/R-94/526. U.S. Environmental Protection Agency, November 1994.

Smith, L.A. and Hinchee, R.E. *In Situ Thermal Technologies for Site Remediation,* Chelsea, MI: Lewis Publishers, 1992.

Trombley, J. "Electrochemical Remediation Takes to the Field." *Environ. Sci. Technol.,* vol. 28, no. 6, pp. 289A–291A, June 1994.

EPA, "Demonstration Program of Hughes Environmental Systems, Inc. (Steam Enhanced Recovery Process)," in *SITE: Technology Profiles*, EPA/540/R-94/526, U.S. Environmental Protection Agency, November 1994a.

EPA, "Demonstration Program of Udell Technologies, Inc. (In Situ Steam Enhanced Extraction Process)," in *SITE: Technology Profiles*, EPA/540/R-94/526, U.S. Environmental Protection Agency, November 1994b.

EPA, "Demonstration Program of Praxis Environmental Technologies, Inc. (In Situ Steam Enhanced Extraction Process)," in *SITE: Technology Profiles*, EPA/540/R-94/526, U.S. Environmental Protection Agency, November 1994c.

EPA, "Demonstration Program of NOVATERRA, Inc. (In Situ Steam and Air Stripping), in *SITE: Technology Profiles*, EPA/540/R-94/526, U.S. Environmental Protection Agency, November 1994d.

EPA, "Demonstration Program of Hrubetz Environmental Services, Inc. (HRUBOUT Process), in *SITE: Technology Profiles*, EPA/540/R-94/526, U.S. Environmental Protection Agency, November 1994e.

EPA, "Emerging Technology Program of Electrokinetics, Inc. (Electrokinetic Remediation)," in *SITE: Technology Profiles*, EPA/540/R-94/526, U.S. Environmental Protection Agency, November 1994f.

EPA, "Emerging Technology Program of Battelle Memorial Institute (In Situ Electroacoustic Soil Decontamination), in *SITE: Technology Profiles*, EPA/540/R-94/526, U.S. Environmental Protection Agency, November 1994g.

EPA, "Demonstration Program of International Waste Technologies/Geo-Con, Inc. (In Situ Solidification and Stabilization Process), in *SITE: Technology Profiles*, EPA/540/R-94/526, U.S. Environmental Protection Agency, November 1994h.

EPA, "Demonstration Program of In Situ Fixation Company (Deep In Situ Bioremediation Process), in *SITE: Technology Profiles*, EPA/540/R-94/526, U.S. Environmental Protection Agency, November 1994i.

EPA, "Demonstration Program of S.M.W. Seiko, Inc. (In Situ Solidification and Stabilization), in *SITE: Technology Profiles*, EPA/540/R-94/526, U.S. Environmental Protection Agency, November 1994j.

EPA, "SITE Demonstration Bulletin: In Situ Vitrification, Geosafe Corporation," EPA/540/MR-94/520, U.S. Environmental Protection Agency, August 1994k.

Index